Dan Evans

ESSENTIALS

AQA
GCSE Chemistry
Revision Guide

Contents

Contents

N.B. The numbers in brackets correspond to the reference numbers
on the AQA GCSE Chemistry specification.

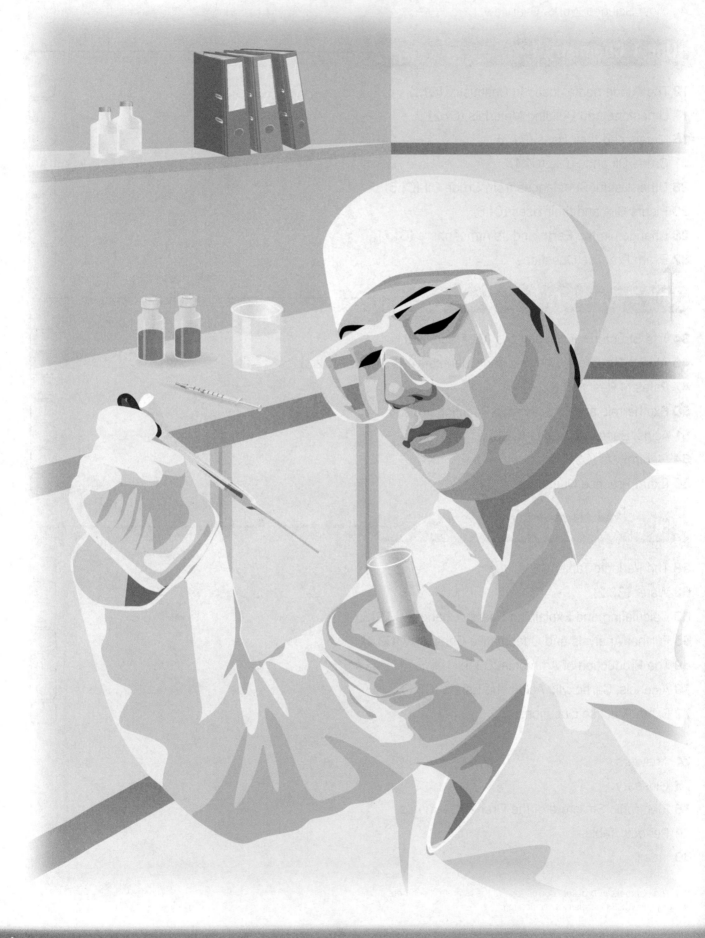

How Science Works – Explanation

The AQA GCSE Chemistry specification incorporates:
- **Science Content** – all the scientific explanations and evidence that you need to know for the exams. (It is covered on pages 12–73 of this revision guide.)
- **How Science Works** – a set of key concepts, relevant to all areas of science. It covers...
  - the relationship between scientific evidence, and scientific explanations and theories
  - how scientific evidence is collected
  - how reliable and valid scientific evidence is
  - the role of science in society
  - the impact science has on our lives
  - how decisions are made about the ways science and technology are used in different situations, and the factors affecting these decisions.

Your teacher(s) will have taught these two types of content together in your science lessons. Likewise, the questions on your exam papers will probably combine elements from both types of content. So, to answer them, you'll need to recall and apply the relevant scientific facts and knowledge of how science works.

The key concepts of How Science Works are summarised in this section of the revision guide (pages 5–11). You should be familiar with all of these concepts. If there is anything you are unsure about, ask your teacher to explain it to you.

How Science Works is designed to help you learn about and understand the practical side of science. It aims to help you develop your skills when it comes to...
- evaluating information
- developing arguments
- drawing conclusions.

The Thinking Behind Science

Science attempts to explain the world we live in.

Scientists carry out investigations and collect evidence in order to...
- **explain phenomena** (i.e. how and why things happen)
- **solve problems** using evidence.

Scientific knowledge and understanding can lead to the **development of new technologies** (e.g. in medicine and industry), which have a huge impact on **society** and the **environment**.

The Purpose of Evidence

Scientific evidence provides **facts** that help to answer a specific question and either **support** or **disprove** an idea or theory. Evidence is often based on data that has been collected through **observations** and **measurements**.

To allow scientists to reach conclusions, evidence must be...
- **repeatable** – other people should be able to repeat the same process
- **reproducible** – other people should be able to reproduce the same results
- **valid** – it must be repeatable, reproducible and answer the question.

N.B. If data isn't repeatable and reproducible, it can't be valid.

To ensure scientific evidence is repeatable, reproducible and valid, scientists look at ideas relating to...
- observations
- investigations
- measurements
- data presentation
- conclusions and evaluation.

How Science Works Overview

Observations

Most scientific investigations begin with an **observation**. A scientist observes an event or phenomenon and decides to find out more about how and why it happens.

The first step is to develop a **hypothesis**, which suggests an explanation for the phenomenon. Hypotheses normally suggest a relationship between two or more **variables** (factors that change).

Hypotheses are based on…

- careful observations
- existing scientific knowledge
- some creative thinking.

The hypothesis is used to make a **prediction**, which can be tested through scientific investigation. The data collected from the investigation will…

- support the hypothesis **or**
- show it to be untrue (refute it) **or**
- lead to the modification of the original hypothesis or the development of a new hypothesis.

If the hypothesis and models we have available to us do not completely match our data or observations, we need to check the validity of our observations or data, or amend the models.

Sometimes, if the new observations and data are valid, existing theories and explanations have to be revised or amended, and so scientific knowledge grows and develops.

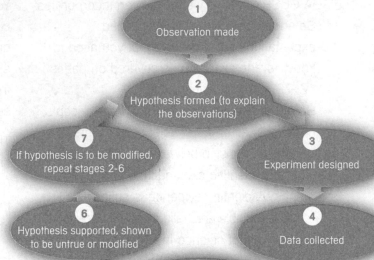

Example

- Two scientists **observe** that freshwater shrimp are only found in certain parts of a stream.
- They use scientific knowledge of shrimp and water flow to develop a **hypothesis**, which relates the presence of shrimp (dependent variable) to the rate of water flow (independent variable). For example, a hypothesis could be: the faster the water flows, the fewer shrimp are found.
- They **predict** that shrimp are only found in parts of the stream where the water flow rate is below a certain value.
- They **investigate** by counting and recording the number of shrimp in different parts of the stream, where water flow rates differ.
- The **data** shows that more shrimp are present in parts of the stream where the flow rate is below a certain value. So, the data **supports** the hypothesis. But, it also shows that shrimp aren't always present in these parts of the stream.
- The scientists realise there must be another factor affecting the distribution of shrimp. They **refine their hypothesis**.

Investigations

An **investigation** involves collecting data to find out whether there is a relationship between two **variables**. A variable is a factor that can take different values.

In an investigation there are two types of variables:

- **Independent** variable – can be changed by the person carrying out the investigation. For example, the amount of water a plant receives.
- **Dependent** variable – measured each time a change is made to the independent variable, to see if it also changes. For example, the growth of the plant (measured by recording the number of leaves).

For a measurement to be valid it must measure only the appropriate variable.

Variables can have different types of values:

- **Continuous variables** – can take any numerical value (including decimals). These are usually measurements, e.g. temperature.
- **Categoric variables** – a variable described by a label, usually a word, e.g. different breeds of dog or blood group.
  - **Discrete variables** – only take whole-number values. These are usually quantities, e.g. the number of shrimp in a stream.
  - **Ordered variables** – have relative values, e.g. 'small', 'medium' or 'large'.

N.B. Numerical values, such as continuous variables, tend to be more informative than ordered and categoric variables.

An investigation tries to find out whether an **observed** link between two variables is...

- **causal** – a change in one variable causes a change in the other, e.g. the more cigarettes you smoke, the greater the chance that you will develop lung cancer.
- **due to association** – the changes in the two variables are linked by a third variable, e.g. as grassland decreases, the number of predators decreases (caused by a third variable, i.e. the number of prey decreasing).
- **due to chance** – the change in the two variables is unrelated; it is coincidental, e.g. people who eat more cheese than others watch more television.

Controlling Variables

In a **fair test**, the only factor that should affect the dependent variable is the independent variable. Other **outside variables** that could influence the results are kept the same, i.e. constant (control variables) or eliminated.

It's a lot easier to control all the other variables in a laboratory than in the field, where conditions can't always be controlled. The impact of an outside variable (e.g. light intensity or rainfall) has to be reduced by ensuring all the measurements are affected by it in the same way. For example, all the measurements should be taken at the same time of day.

Control groups are often used in biological and medical research to make sure that any observed results are due to changes in the independent variable only.

A sample is chosen that 'matches' the test group as closely as possible except for the variable that is being investigated, e.g. testing the effect of a drug on reducing blood pressure. The control group must be the same age, gender, have similar diets, lifestyles, blood pressure, general health, etc.

Investigations (Cont.)

Accuracy and Precision

How accurate data needs to be depends on what the investigation is trying to find out. For example, when measuring the volume of acid needed to neutralise an alkaline solution it is important that equipment is used that is able to accurately measure volumes of liquids.

The data collected must be **precise** enough to form a **valid conclusion**: it should provide clear evidence for or against the hypothesis.

To ensure data is as accurate as possible, you can…

- calculate the **mean** (average) of a set of repeated measurements to reduce the effect of random errors
- increase the number of measurements taken to improve the reliability of the mean / spot anomalies.

Preliminary Investigations

A trial run of an investigation will help identify appropriate values to be recorded, such as the number of repeated readings needed and their range and interval.

Measurements

Apart from control variables, there are a number of factors that can affect the reliability and validity of measurements:

- **Accuracy of instruments** – depends on how accurately the instrument has been calibrated. An accurate measurement is one that is close to the true value.
- **Resolution (or sensitivity) of instruments** – determined by the smallest change in value that the instrument can detect. The more sensitive the instrument, the more **precise** the value. For example, bathroom scales aren't sensitive enough to detect changes in a baby's mass, but the scales used by a midwife are.
- **Human error** – even if an instrument is used correctly, human error can produce random differences in repeated readings or a systematic shift from the true value if you lose concentration or make the same mistake repeatedly.
- **Systematic error** – can result from repeatedly carrying out the process incorrectly, making the same mistake each time.
- **Random error** – can result from carrying out a process incorrectly on odd occasions or by fluctuations in a reading. The smaller the random error the greater the accuracy of the reading.

You need to examine any **anomalous** (irregular) values to try to determine why they appear. If they have been caused by equipment failure or human error, it is common practice to ignore them and not use them in any calculations.

There will always be some variation in the actual value of a variable, no matter how hard we try to repeat an event.

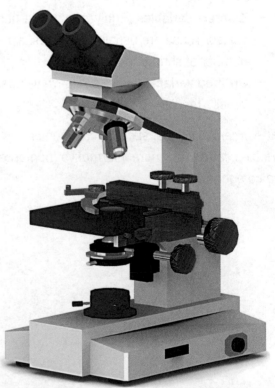

Presenting Data

Data is often presented in a **chart** or **graph** because it makes...

- any patterns more obvious
- it easier to see the relationship between two variables.

The **mean** (or average) of data is calculated by adding all the measurements together, then dividing by the number of measurements taken:

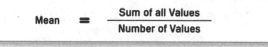

$$\text{Mean} = \frac{\text{Sum of all Values}}{\text{Number of Values}}$$

If you present data clearly, it is easier to identify any anomalous (irregular) values. The type of chart or graph you use to present data depends on the type of variable involved:

1 **Tables** organise data (but patterns and anomalies aren't always obvious)

Height of student (cm)	127	165	149	147	155	161	154	138	145
Shoe size	5	8	5	6	5	5	6	4	5

2 **Bar charts** display data when the independent variable is categoric or discrete and the dependent variable is continuous.

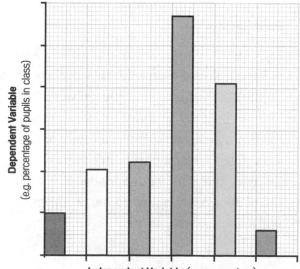

3 **Line graphs** display data when both variables are continuous.

- Points are joined by straight lines if you don't have data to support the values between the points.
- A line of best fit is drawn if there is sufficient data or if a trend can be assumed.

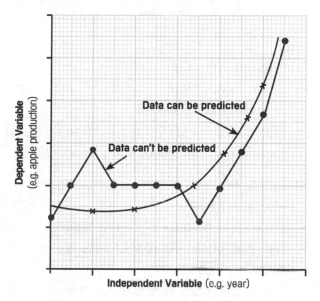

4 **Scattergrams** (scatter diagrams) show the underlying relationship between two variables. This can be made clearer if you include a **line of best fit**. A line of best fit could be a straight line or a smooth curve.

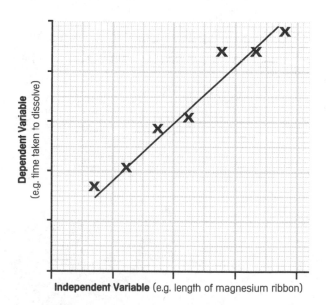

Conclusions and Evaluations

Conclusions **should**…
- describe patterns and relationships between variables
- take all the data into account
- make direct reference to the original hypothesis or prediction
- try to explain the results / observations by making reference to the hypothesis as appropriate.

Conclusions **should not**…
- be influenced by anything other than the data collected (i.e. be biased)
- disregard any data (except anomalous values)
- include any unreasoned speculation.

An **evaluation** looks at the whole investigation. It should consider…
- the original purpose of the investigation
- the appropriateness of the methods and techniques used
- the reliability and validity of the data
- the validity of the conclusions.

The **reliability** of an investigation can be increased by…
- looking at relevant data from secondary sources (i.e. sources created by someone who did not experience first hand or participate in the original experiment)
- using an alternative method to check results
- ensuring results can be reproduced by others.

Science and Society

Scientific understanding can lead to technological developments. These developments can be exploited by different groups of people for different reasons. For example, the successful development of a new drug…
- benefits the drugs company financially
- improves the quality of life for patients
- can benefit society (e.g. if a new drug works, then maybe fewer people will be in hospital, which reduces time off sick, cost to the NHS, etc).

Scientific developments can raise certain **issues**. An issue is an important question that is in dispute and needs to be settled. The resolution of an issue may not be based on scientific evidence alone.

There are several different types of **issue** that can arise:
- **Social** – the impact on the human population of a community, city, country, or the world.
- **Economic** – money and related factors like employment and the distribution of resources.
- **Environmental** – the impact on the planet, its natural ecosystems and resources.
- **Ethical** – what is morally right or wrong; requires a value judgement to be made.

N.B. There is often an overlap between social and economic issues.

Peer Review

Finally, peer review is a process of self-regulation involving qualified professional individuals or experts in a particular field who examine the work undertaken critically. The vast majority of peer review methods are designed to maintain standards and provide credibility for the work that has been undertaken. These methods vary depending on the nature of the work and also on the overall purpose behind the review process.

Evaluating Information

It is important to be able to evaluate information relating to social-scientific issues, for both your GCSE course and to help you make informed decisions in life.

When evaluating information...
- make a list of **pluses** (pros)
- make a list of **minuses** (cons)
- consider how each point might **impact on society**.

You also need to consider whether the source of information is reliable and credible. Some important factors to consider are...
- **opinions**
- **bias**
- **weight of evidence**.

Opinions are personal viewpoints. Opinions backed up by valid and reliable evidence carry far more weight than those based on non-scientific ideas.

Opinions of experts can also carry more weight than non-experts.

Information is **biased** if it favours one particular viewpoint without providing a balanced account.

Biased information might include incomplete evidence or try to influence how you interpret the evidence.

Scientific evidence can be given **undue weight** or dismissed too quickly due to...
- political significance (consequences of the evidence could provoke public or political unrest)
- status of the experiment (e.g. if they do not have academic or professional status, experience, authority or reputation).

Limitations of Science

Although science can help us in lots of ways, it can't supply all the answers. We are still finding out about things and developing our scientific knowledge.

There are some questions that science can't answer. These tend to be questions...
- where beliefs, opinions and ethics are important
- where we don't have enough reproducible, repeatable or valid evidence.

Science can often tell us if something **can** be done, and **how** it can be done, but it can't tell us whether it **should** be done.

Decisions are made by individuals and by society on issues relating to science and technology.

C1 The Fundamental Ideas in Chemistry

Atoms and Elements

All substances are made of **atoms** (very small particles). Each atom has a small central **nucleus** made up of **protons** and **neutrons**. The nucleus is surrounded by orbiting **electrons**.

A substance that contains only one sort of atom is called an **element**. There are about 100 different elements.

The atoms of each element are represented by a different **chemical symbol**.

For example…
- sodium = Na
- carbon = C
- iron = Fe

An Atom

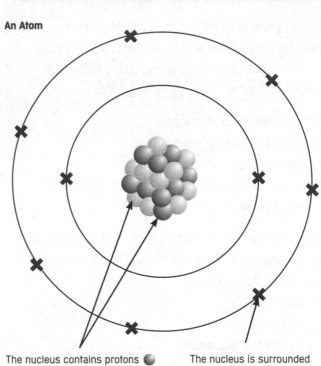

The nucleus contains protons and neutrons ●

The nucleus is surrounded by orbiting electrons ✘

The Periodic Table

Elements are arranged in the **Periodic Table**. The **groups** in the Periodic Table contain elements that have similar properties.

For example, all Group 1 elements (the alkali metals) react vigorously with water to produce an alkaline solution and hydrogen gas.

The Group 1 elements react rapidly with oxygen to form metal oxides. The elements in Group 0 are called the noble gases. They are all unreactive elements because their atoms have full outer shells of electrons, meaning they are stable. Atoms of noble gases have eight electrons in their outer shell, except for helium, which only has two.

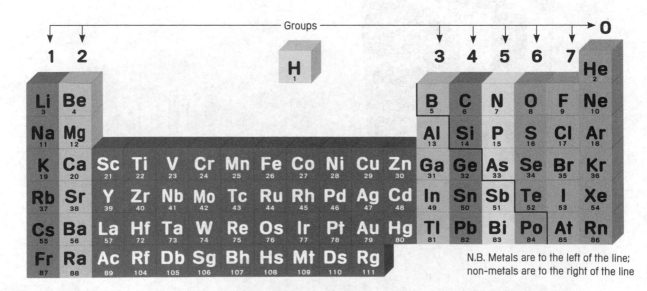

N.B. Metals are to the left of the line; non-metals are to the right of the line

Key Words   **Atom • Nucleus • Proton • Neutron • Electron • Element**

The Fundamental Ideas in Chemistry C1

Atomic Structure

Protons, neutrons and electrons have relative **electrical charges**.

Atomic Particle	Relative Charge
Proton ●	+1
Neutron ●	0
Electron ✖	-1

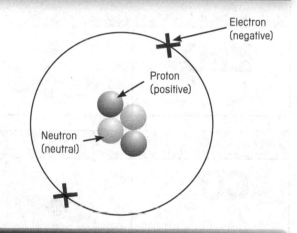

Electron (negative)

Proton (positive)

Neutron (neutral)

Atoms, as a whole, have **no overall charge** because they contain an **equal number** of **protons** and **electrons**.

All atoms of a particular element have the **same number of protons**. Atoms of different elements have **different numbers** of **protons**. The number of protons in an atom is called its **atomic number**. The sum of the protons and neutrons in an atom is its **mass number**.

Elements are arranged in the modern Periodic Table in order of **atomic number**.

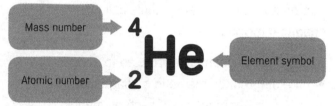

Mass number → 4

Atomic number → 2 He ← Element symbol

The number of neutrons in an atom is calculated by:

Mass number — Atomic number

Electron Configuration and Structure

Electrons in an atom occupy the lowest available **energy level** (i.e. the innermost available shell). For the first 20 elements:

- The **first** level can only contain a **maximum** of **two electrons**.
- The energy levels after this can each hold a **maximum** of **eight electrons**.

The **electron configuration** tells us how the electrons are arranged around the nucleus in **energy levels** or **shells**. It is written as a series of numbers, for example...

- oxygen is 2,6
- aluminium is 2,8,3

The **Periodic Table** groups the elements in terms of **electronic structure**.

Elements in the **same group** have the same number of **electrons** in their **highest energy level (outer shell)**

and this gives them **similar chemical properties**. A particular energy level is gradually filled with electrons from **left to right**, across each **period**.

Electronic configurations can also be represented as shown below.

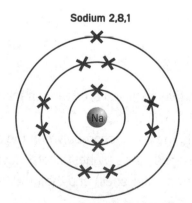

Sodium 2,8,1

You need to be able to represent the electronic structure of the first 20 elements (see page 78).

C1 The Fundamental Ideas in Chemistry

Compounds

When elements react their atoms join with other atoms to form **compounds**. This involves…

- the **giving** and **taking** of electrons to form ions **OR**
- the **sharing** of electrons to form **molecules**.

Metal atoms lose electrons to form positive ions and non-metal atoms gain electrons to form negative ions. Compounds formed between metals and non-metals consist of ions. Compounds formed between non-metal atoms form molecules. Atoms in molecules are held together by covalent bonds.

Chemical Formulae

Compounds are represented by a combination of numbers and chemical symbols called a **chemical formula**.

Scientists use **chemical formulae** to show…
- the different elements in a compound
- the number of atoms of each element in the compound.

In chemical formulae, the position of the numbers tells you what is multiplied:
- A small number that sits below the line multiplies only the symbol that comes immediately before it.
- A number that is the same size as the letters multiplies all the symbols that come after it.

For example…
- H_2O means $(2 \times H) + (1 \times O)$
- $2NaOH$ means $2 \times (NaOH)$ or $2 \times (Na + O + H)$.

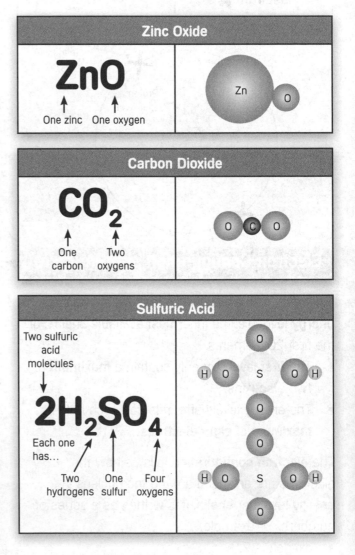

Zinc Oxide

ZnO
One zinc   One oxygen

Carbon Dioxide

CO_2
One carbon   Two oxygens

Sulfuric Acid

Two sulfuric acid molecules
$2H_2SO_4$
Each one has…
Two hydrogens   One sulfur   Four oxygens

Quick Test

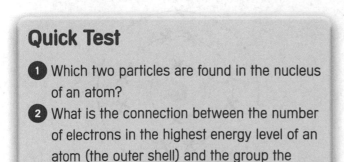

1. Which two particles are found in the nucleus of an atom?
2. What is the connection between the number of electrons in the highest energy level of an atom (the outer shell) and the group the element is in?
3. What is the electron configuration of an atom of magnesium?

Chemical Reactions

You can show what has happened during a **chemical reaction** by writing a **word** or **symbol equation**.

The **reactants** (i.e. the substances that react) are on one side of the equation and the **products** (i.e. the new substances that are formed) are on the other.

The total mass of the products of a chemical reaction is always equal to the total mass of the reactants. This is because **no atoms are lost or made**. The products of a chemical reaction are made up from exactly the same atoms as the reactants.

Chemical symbol equations must always be **balanced**. There must be the same number of atoms of each element on the reactant side of the equation as there is on the product side.

Number of atoms of each element on reactants side	=	Number of atoms of each element on products side

Example

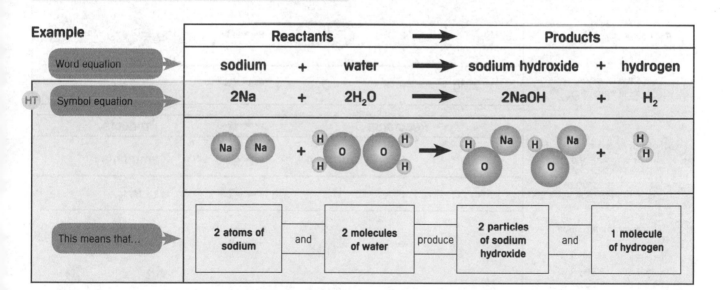

	Reactants	→	Products
Word equation	sodium + water	→	sodium hydroxide + hydrogen
Symbol equation (HT)	$2Na$ + $2H_2O$	→	$2NaOH$ + H_2

This means that...

2 atoms of sodium	and	2 molecules of water	produce	2 particles of sodium hydroxide	and	1 molecule of hydrogen

HT Writing Balanced Equations

The following steps tell you how to write a balanced equation.

1. Write a word equation for the chemical reaction.

2. Substitute formulae for the elements or compounds.

3. Balance the equation by adding numbers in front of the reactants and / or products.

4. Write down the balanced symbol equation.

	Reactants	→	Products
1 Write a word equation	magnesium + oxygen	→	magnesium oxide
2 Substitute formulae	Mg + O_2	→	MgO

3 Balance the equation

- There are two **O**s on the reactant side, but only one **O** on the product side. We need to add another **MgO** to the product side to balance the **O**s.
- We now need to add another **Mg** on the reactant side to balance the **Mg**s.
- There are two magnesium atoms and two oxygen atoms on each side – **it is balanced.**

	Reactants	→	Products
4 Write a balanced symbol equation	$2Mg$ + O_2	→	$2MgO$

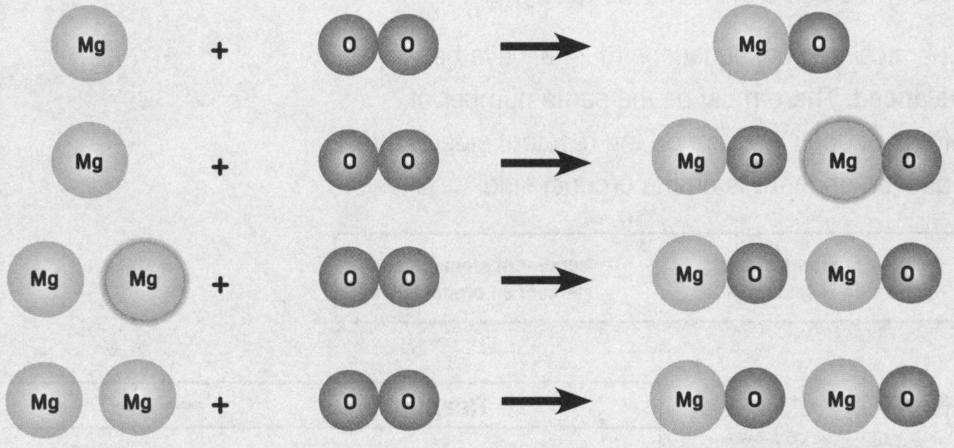

	Reactants	→	Products
1 Write a word equation	nitrogen + hydrogen	→	ammonia
2 Substitute formulae	N_2 + H_2	→	NH_3

3 Balance the equation

- There are two **N**s on the reactant side, but only one **N** on the product side. We need to add another **NH₃** to the product side to balance the **N**s
- We now need to add two more **H₂**s on the reactant side to balance the **H₂**s
- There are two nitrogen atoms and six hydrogen atoms on each side – **it is balanced.**

	Reactants	→	Products
4 Write a balanced symbol equation	N_2 + $3H_2$	→	$2NH_3$

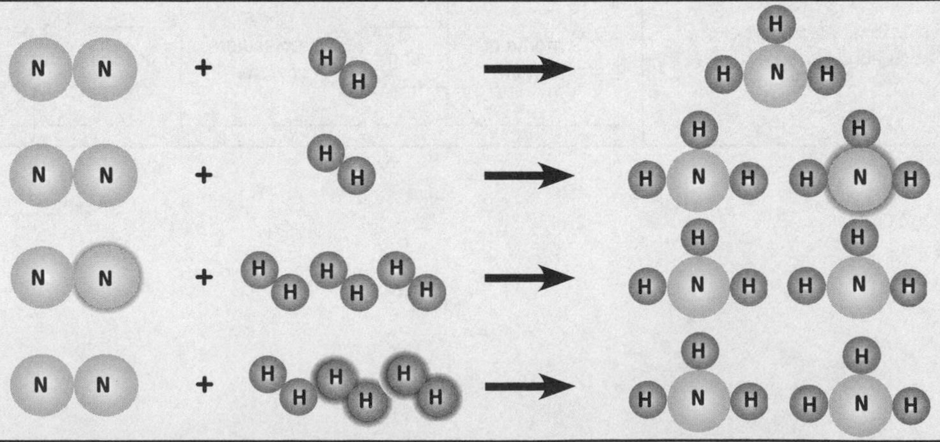

Limestone (CaCO₃)

Limestone is a **sedimentary rock** consisting mainly of **calcium carbonate**. It is cheap, easy to obtain and has many uses.

Limestone can be used…
- as a building material
- for making cement, mortar and concrete.

As a building material

Limestone can be quarried, cut into blocks and used to build houses. It can be **eroded** by **acid rain** but this is a very slow process.

Heating limestone

When calcium carbonate is heated in a kiln it **decomposes**. This reaction is called **thermal decomposition**. It causes the calcium carbonate to break down into calcium oxide and carbon dioxide.

Magnesium, copper, zinc, calcium and sodium carbonates decompose on heating in a similar way.

Some metal carbonates, e.g. others in Group 1, may not decompose at the temperatures reached by a Bunsen burner.

The calcium oxide can then be reacted with water to produce calcium hydroxide. Calcium hydroxide can be used to neutralise soils and lakes, preventing crop failure.

Carbonates of other metals decompose in a similar way when they're heated.

Making cement, mortar and concrete

Powdered limestone is roasted in a rotary kiln with powdered clay to produce dry **cement**. When sand and water are mixed in, **mortar** is produced.

Mortar is used to hold bricks and stones together. When aggregate, sand and water are mixed in, **concrete** is produced.

Reacting Carbonates with Dilute Acid

Carbonates react with **dilute acids** to form **carbon dioxide** gas (and a salt and water). Carbon dioxide turns limewater (a solution of calcium hydroxide in water) cloudy. For example…

calcium carbonate	+	hydrochloric acid	→	calcium chloride	+	carbon dioxide	+ water

$$CaCO_3(s) + 2HCl(aq) \rightarrow CaCl_2(aq) + CO_2(g) + H_2O(l)$$

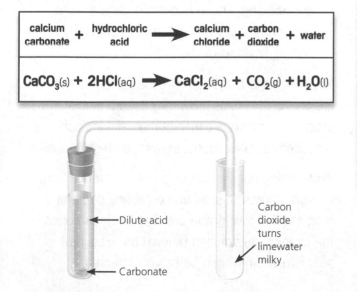

Dilute acid

Carbonate

Carbon dioxide turns limewater milky

Quick Test

1. Calcium carbonate can be decomposed upon heating (thermal decomposition) to make which two compounds?
2. What is the chemical test for carbon dioxide gas?
3. What substance is powdered limestone mixed with to make cement?
4. **HT** Balance the following equations:
   a) $Al + O_2 \longrightarrow Al_2O_3$
   b) $Cr + HCl \longrightarrow CrCl_3 + H_2$

C1 Metals and their Uses

Ores

The Earth's crust contains many naturally occurring **elements** and **compounds** called **minerals**.

A **metal ore** is a mineral that contains enough metal to make it economically viable to extract it. Over time it may become more or less economic to extract a metal from an ore as technology develops.

Ores are mined and impurities may be removed before the metal is extracted and purified.

This mining can involve the digging up and processing of large amounts of rock.

Extracting Metals from their Ores

The method of metal extraction depends on **how reactive the metal is**.

Unreactive metals, like gold, exist naturally in the earth and can be obtained through panning. But most metals are found as **metal oxides**, or as compounds that can be easily changed into a metal oxide.

Metals that are **less reactive than carbon** can be extracted from their oxides by heating with carbon, e.g. iron and lead.

Metals **more reactive than carbon**, e.g. aluminium, are extracted by electrolysis of molten compounds.

Extraction of Copper

Copper is a useful metal because it is a good conductor of electricity and heat. It is easily bent into shape yet hard enough to be used to make water pipes and tanks. It does not react with water so lasts for a long time.

Copper can be extracted from copper-rich ores by heating the ores in a furnace. This process is known as **smelting**. This copper can then be purified by **electrolysis**. Copper can also be obtained from solutions of copper salts by electrolysis or by displacement using scrap iron.

During electrolysis the positive copper ions move towards the negative electrode and form pure copper metal.

But the mining of more copper means that we are running out of copper-rich ores. So, new methods have been developed to extract copper from ores that contain less copper.

Copper can be extracted from:
- Low-grade ores (ores that contain small amounts of copper)
- Contaminated land by **phytomining** or by **bioleaching**. These two methods are more environmentally friendly than traditional mining methods.

Phytomining is a method that uses plants to absorb copper. As the plants grow they absorb (and store) copper. The plants are then burned and the ash produced contains copper in relatively high quantities.

Bioleaching uses bacteria to extract metals from low-grade ores. A solution containing bacteria is mixed with a low-grade ore. The bacteria convert the copper into solution (known as a leachate solution) where it can be easily extracted.

Key Words   **Mineral • Ore • Smelting • Electrolysis • Phytomining • Bioleaching**

Iron

Iron oxide can be reduced in a blast furnace to produce **iron**. Molten iron obtained from a blast furnace contains roughly…

- 96% iron
- 4% carbon and other metals.

Because it is impure the iron is very brittle, with limited uses. To produce pure iron, all the **impurities** have to be removed.

The **atoms** in pure iron are arranged in layers that can slide over each other easily. This makes pure iron soft and malleable. It can be easily shaped, but it's too soft for many practical uses.

The properties of iron can be changed by mixing it with small quantities of carbon or other metals to make **steel**. The majority of iron is converted into steel. Steel is an **alloy**.

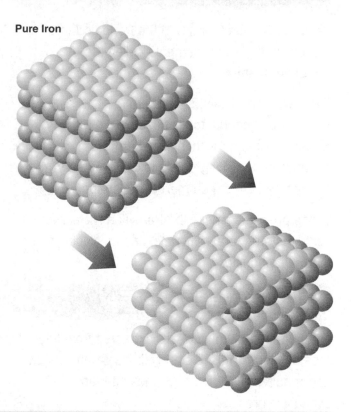

Pure Iron

Alloys

Many of the metals you come across every day are alloys. Pure copper, gold and aluminium are too soft for many uses, so they are mixed with small amounts of similar metals to make them harder for items in everyday use, for example coins.

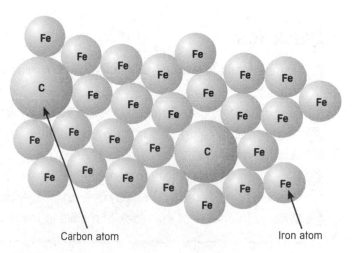

Steel

Carbon is added to iron to make the alloy **steel**.

Alloys like steel are developed to have the necessary properties for a specific purpose.

In steel, the amount of carbon and / or other elements determines its properties:

- Steel with a high carbon content is hard and strong.
- Steel with a low carbon content is soft and easily shaped.
- **Stainless steel** is hard and resistant to corrosion.

Carbon atom

Iron atom

C1 Metals and their Uses

The Transition Metals

Between Groups 2 and 3 in the Periodic Table is a block of metallic elements called the **transition metals**.

The transition metals…
- are good **conductors** of heat and electricity
- are hard and mechanically strong
- have high melting points (except mercury)
- can be bent or hammered into shape.

These properties make the transition metals very useful as structural materials, and as electrical and thermal conductors.

2					Transition metals					3
Sc 21	Ti 22	V 23	Cr 24	Mn 25	Fe 26	Co 27	Ni 28	Cu 29	Zn 30	
Y 39	Zr 40	Nb 41	Mo 42	Tc 43	Ru 44	Rh 45	Pd 46	Ag 47	Cd 48	
La 57	Hf 72	Ta 73	W 74	Re 75	Os 76	Ir 77	Pt 78	Au 79	Hg 80	
Ac 89	Rf 104	Db 105	Sg 106	Bh 107	Hs 108	Mt 109	Ds 110	Rg 111		

Extracting Metals

Titanium and **aluminium** are extracted from their ores by **electrolysis**. Electrolysis has many stages and requires a lot of energy, making it an expensive process.

Copper is useful for chemical wiring and plumbing.

Aluminium is resistant to corrosion and has a low density so it's very light.

Aluminium is used for drinks cans, window frames, lightweight vehicles and aeroplanes.

Titanium is strong and resistant to corrosion. It is used in aeroplanes, nuclear reactors and replacement hip joints.

Recycling Metals

Metals should be **recycled** wherever possible to…
- save money and energy
- make sure natural resources aren't used up
- reduce damage to the environment.

Quick Test

1. Complete the following sentence: Ores are mined and may be _____ before the metal is extracted and then _____ .
2. Iron is converted into steel by mixing iron with which element?
3. Copper can be extracted from low-grade ores by what two methods?
4. What is an alloy?
5. Why is pure iron soft and malleable?
6. Why are most metals in everyday use, e.g. copper and aluminium, mixed with small amounts of other metals?
7. What is the name of the block of metals in the middle of the Periodic Table?

Crude Oil

Crude oil on its own isn't very useful. But it is a **mixture** of a large number of compounds, some of which are very useful. Crude oil is a limited resource that is used to produce fuels and other chemicals.

A **mixture** consists of two or more elements or compounds that are **not chemically combined** together. The properties of the substances in a mixture remain unchanged, so they can be separated by physical methods, such as **distillation**.

Most of the compounds in crude oil consist of molecules made up of only **carbon** and **hydrogen** atoms. These compounds are called **hydrocarbons**. **Hydrocarbon** molecules vary in size, which affects their properties and how they are used as fuels.

The larger the hydrocarbon (i.e. the more carbon and hydrogen atoms in a molecule)...

- the less easily it flows (it's more viscous)
- the higher its boiling point
- the less volatile it is
- the less easily it ignites.

Long-chain Hydrocarbon   **Short-chain Hydrocarbons**

Fractional Distillation

Crude oil can be separated into different **fractions** (parts) by **fractional distillation.**

Each fraction contains hydrocarbon molecules with a similar number of carbon atoms. Most of the hydrocarbons obtained are **alkanes** (**saturated hydrocarbons**).

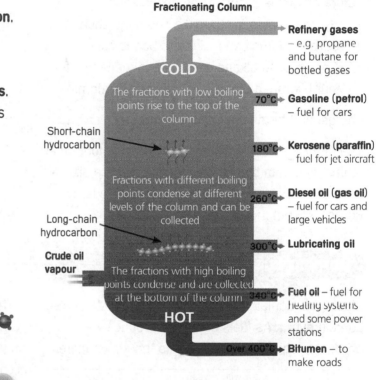

Fractionating Column

COLD

The fractions with low boiling points rise to the top of the column

Short-chain hydrocarbon

Fractions with different boiling points condense at different levels of the column and can be collected

Long-chain hydrocarbon

Crude oil vapour

The fractions with high boiling points condense and are collected at the bottom of the column

HOT

Refinery gases – e.g. propane and butane for bottled gases

70°C **Gasoline (petrol)** – fuel for cars

180°C **Kerosene (paraffin)** – fuel for jet aircraft

260°C **Diesel oil (gas oil)** – fuel for cars and large vehicles

300°C **Lubricating oil**

340°C **Fuel oil** – fuel for heating systems and some power stations

Over 400°C **Bitumen** – to make roads

Ethanol and Hydrogen as Fuels

There are advantages and disadvantages to using ethanol and hydrogen as fuels.

Hydrogen as a biofuel	
Advantages	**Disdvantages**
<ul><li>Water is the only product of combustion so it's a 'clean' fuel</li><li>Water is potentially a source of plentiful supplies of hydrogen</li></ul>	<ul><li>There are currently no 'low energy' ways to extract hydrogen from water in large quantities</li><li>Hydrogen is a gas, so is difficult to store in large quantities</li><li>When hydrogen is mixed with air and ignited it's explosive, so there are safety issues to consider</li></ul>

Ethanol as a biofuel	
Advantages	**Disdvantages**
<ul><li>A renewable energy source, i.e. helps to preserve fossil fuels</li><li>Sugar cane/beet (used to produce ethanol) grows quickly in countries with a hot climate, e.g. Brazil</li><li>Sugar cane/beet absorbs CO_2 from the atmosphere as it grows</li></ul>	<ul><li>Sugar cane/beet can only be grown in countries with a hot climate</li><li>CO_2 is produced as a product of combustion (CO_2 is a greenhouse gas)</li></ul>

C1 Crude Oil and Fuels

Biofuels

Biofuels, e.g. biodiesel and ethanol, are fuels that are produced from plant material such as sugar.

The production and use of biofuels often have environmental advantages over traditional fuels.

Alkanes (Saturated Hydrocarbons)

The 'spine' of a hydrocarbon is made up of a chain of carbon atoms. When these chains are joined together by **single carbon–carbon bonds** the hydrocarbon is **saturated** and is known as an **alkane**.

- Hydrogen atoms can make one bond each.
- Carbon atoms can make four bonds each.
- The simplest alkane, **methane**, is made up of four hydrogen atoms and one carbon atom.

The general formula for alkanes is C_nH_{2n+2}

The carbon atoms in alkenes are linked to four other atoms by **single bonds**. This means that the alkane is saturated. This explains why alkanes are fairly unreactive, but they do burn well.

The shorter-chain hydrocarbons release energy more quickly by burning, so there is greater demand for them as **fuels**.

Alkanes can be represented as seen below. The '–' between atoms represents a covalent bond.

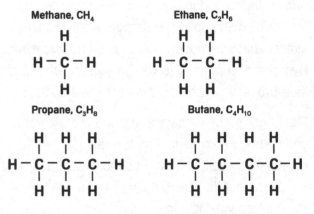

Burning Fuels

Most **fuels** contain carbon and hydrogen. Many also contain **sulfur**. As fuels burn they produce waste products, which are released into the atmosphere.

carbon	burn with oxygen →	carbon dioxide
C + O$_2$	→	CO$_2$

carbon	burn with oxygen →	carbon monoxide
2C + O$_2$	→	2CO

hydrogen	burn with oxygen →	water vapour
2H$_2$ + O$_2$	→	2H$_2$O

sulfur	burn with oxygen →	sulfur dioxide
S + O$_2$	→	SO$_2$

The combustion of hydrocarbon fuel releases energy. During combustion, both the carbon and hydrogen are oxidised. If combustion is not complete, then solid particles containing soot (carbon) and unburnt fuels may be released.

Due to the high temperature reached when fuels burn, nitrogen in the air can react with oxygen to form nitrogen oxides. These nitrogen oxides, like sulfur dioxide, can cause acid rain.

Carbon dioxide causes **global warming** due to the greenhouse effect. Solid particles in the air cause **global dimming**.

Sulfur can be removed from fuel before burning (e.g. in motor vehicles). Sulfur dioxide can be removed from the waste gases after **combustion** (e.g. in power stations). But both of these processes add to costs.

Biofuel • Saturated • Alkane • Fuel

Cracking Hydrocarbons

Longer-chain hydrocarbons can be broken down into shorter, more useful hydrocarbons. This process is called **cracking**.

Long-chain hydrocarbon → (heat + catalyst) → Short-chain hydrocarbons

Cracking involves...
- heating the hydrocarbons until they vaporise
- passing the vapour over a hot **catalyst** (or mixed with steam).

A **thermal decomposition** reaction then takes place.

The products of cracking include **alkanes** and **alkenes (unsaturated hydrocarbons)**. Some of the products are useful as fuels. Alkenes react with bromine water, turning it from orange to colourless.

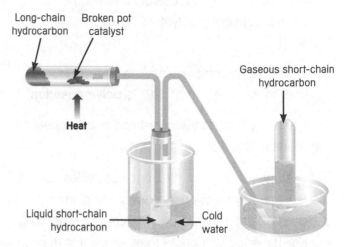

Cracking Hydrocarbons in the Laboratory

Alkenes (Unsaturated Hydrocarbons)

As well as forming single bonds with other atoms, carbon atoms can form **double bonds**. This means that not all the carbon atoms are linked to four other atoms; a **double carbon–carbon bond** is present instead.

Some of the products of cracking are hydrocarbon molecules with at least one double bond (**alkenes**).
- The general formula for alkenes is C_nH_{2n}
- The simplest alkene is ethene, C_2H_4

- Ethene is made up of four hydrogen atoms and two carbon atoms, and contains one double carbon–carbon bond.

Alkenes can be represented using displayed formulae:

Ethene, C_2H_4 — Double bond

Propene, C_3H_6

Making Alcohol from Ethene

Ethanol is an alcohol. It can be produced by reacting the alkene **ethene** with steam in the presence of a **catalyst**, phosphoric acid.

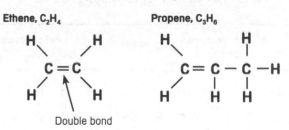

ethene + steam → (phosphoric acid) → ethanol

Making Alcohol by Fermentation

Ethanol can be produced by the **fermentation** of sugar, which is a renewable resource. During fermentation, sugar is converted into ethanol and carbon dioxide.

sugar → ethanol + carbon dioxide

C1 Other Useful Substances from Crude Oil

Polymerisation

Because alkenes are unsaturated (i.e. they contain a double bond), they are useful for making other molecules, especially **polymers** (long-chain molecules).

Many small alkene molecules (monomers) join together to form polymers. This is **polymerisation**.

Polymers such as **poly(ethene)** and **poly(propene)** are made in this way.

For example, slime with different **viscosities** can be made from poly(ethenol). The viscosity of the slime depends on the temperature and concentrations of the poly(ethenol) and borax from which it is made.

The materials commonly called **plastics** are all synthetic polymers.

The small alkene molecules are called monomers.

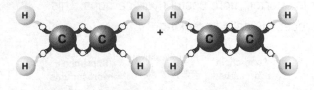

Their double bonds are easily broken.

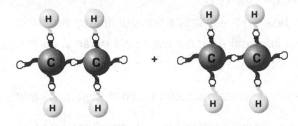

Large numbers of molecules can therefore be joined in this way.

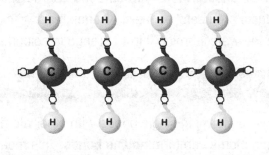

Representing Polymerisation

Polymerisation can be represented like this:

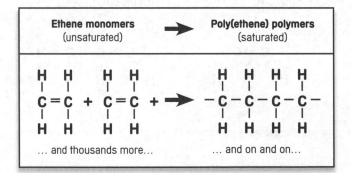

| Ethene monomers (unsaturated) | → | Poly(ethene) polymers (saturated) |

... and thousands more...　　... and on and on...

The general formula for polymerisation can be used to represent the formation of any simple polymer:

$$n\left[C=C \right] \rightarrow \left[-C-C- \right]_n$$

where *n* is a very large number

Biodegradable Polymers

Many polymers are not **biodegradable**, so they are not broken down by microbes. This can lead to problems with waste disposal. Plastic bags made from polymers and cornstarch are being produced so that they are biodegradable.

Key Words　　　　**Polymer • Polymerisation • Biodegradable**

Polymers

Polymers have many useful applications and new uses are being developed. Polymers and composites are widely used in medicine and dentistry. For example…

- implantable materials are used for tissue surgery
- hard-wearing anti-bacterial dental cements, coating and fillers are used in dentistry
- hydrogels can be used as wound dressings.

Polymers can also be used to coat fabrics with a waterproof layer.

Smart materials, including shape-memory polymers, are also increasingly more common.

Specific polymers can have different uses, for example…

- polyvinyl chloride (PVC) is used to make waterproof items and drain pipes, and can be used as an electrical insulator
- polystyrene is used to make the casing for electrical appliances, and it can be expanded to make protective packaging
- poly(ethene) is used to make plastic bags and bottles
- poly(propene) is used to make crates and ropes.

Disposing of Plastics

Plastic is a versatile material. It's cheap and easy to produce, but this means a lot of plastic waste is generated.

There are various ways of **disposing of plastics**. Unfortunately some of them have an impact on the environment.

Burning plastics produces air **pollution**. Burning releases carbon dioxide, which contributes to **global warming**.

Some plastics can't be burned at all because they produce toxic fumes.

Plastics can be dumped in **landfill sites**. But most plastics are **non-biodegradable**. This means that microorganisms have no effect on them, so they will not **decompose** and rot away. The use of landfill sites means that plastic waste builds up.

Research is currently being carried out on the development of **biodegradable plastics**.

Quick Test

1. Which two chemical elements make up most of the molecules in crude oil?
2. What is the name of the process by which crude oil is separated?
3. Name two gases that may be released into the atmosphere when a fuel burns.
4. What are the major environmental concerns over releasing carbon dioxide and sulfur dioxide into the atmosphere?
5. What name is given to the process by which hydrocarbons are broken down into smaller, more useful molecules?
6. What name is given to unsaturated hydrocarbons?
7. What is the chemical test for alkenes?
8. What name is given to the small molecules that join together to form polymers?

C1 Plant Oils and their Uses

Getting Oil from Plants

Many plants produce fruit, seeds and nuts that are rich in **oils.**

The oil can be extracted from plant materials by pressing (crushing) them or by **distillation**. These processes remove the water and other impurities from the plant material.

Vegetable Oils

Vegetable oils are important foods. These oils provide you with nutrients and energy. Vegetable oils can also be used as a **fuel** in converted vehicles, instead of petrol or diesel.

Vegetable oils contain **double carbon–carbon bonds**, so they are **unsaturated**. They can be detected using **bromine water**. Unsaturated fats (oils) **decolourise bromine water**. Vegetable oils are used in cooking because they have a higher boiling point than water so can be used to cook foods at higher temperatures. This means that food can be cooked more quickly and a different flavour is added to food.

Cooking using vegetable oil also increases the energy that food releases when it is eaten.

Cooking with oils higher in unsaturated fats is believed to be healthier than cooking with saturated fats.

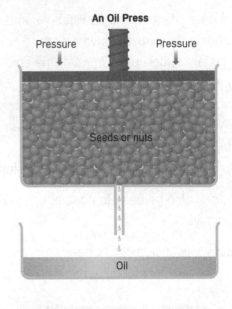

An Oil Press

Pressure   Pressure

Seeds or nuts

Oil

HT Use of Vegetable Oils in Cooking

Generally, the more double carbon–carbon bonds there are in a substance, the lower its melting point.

So, **unsaturated fats** (oils) tend to have melting points below room temperature.

The melting point of an oil can be raised above room temperature by removing some or all of its carbon–carbon bonds. This hardens the oil into a solid fat, for example margarine, which can be spread on bread or used for making cakes and pastries.

This hardening process is called **hydrogenation**.
1. The **unsaturated fat (oil)** is heated with **hydrogen** at about 60°C, in the presence of a **nickel catalyst**.
2. A reaction takes place that **removes** the double carbon–carbon bonds to produce a **saturated fat** (**hydrogenated oil**). Removing more double bonds makes the saturated fat harder.

$$\text{unsaturated fat} + \text{hydrogen} \xrightarrow[\text{catalyst}]{\text{nickel}} \text{saturated fat}$$

Key Words   **Hydrogenation**

Emulsions

Oils don't dissolve in water, but an oil can be mixed with water to produce an **emulsion**.

Emulsions are thicker than oil or water and have a...
- better texture
- better appearance
- better coating ability.

Emulsions have many uses, for example in...
- salad dressings
- ice cream
- cosmetics
- paints

HT Hydrophobic and Hydrophilic Properties of Emulsifiers

An **emulsifier** is a substance that helps to stabilise an emulsion. Emulsifier molecules have a...
- **hydrophilic** (water loving) head that mixes with water molecules
- **hydrophobic** (water hating) tail that mixes with oil molecules.

This allows water and oils to mix.

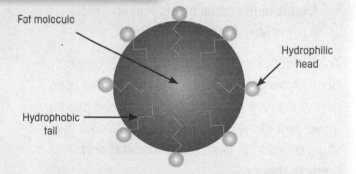

Fat molecule

Hydrophilic head

Hydrophobic tail

Quick Test

1. Why are vegetable oils important foods?
2. Why can vegetable oils be used to cook foods at higher temperatures than water?
3. Give two uses of emulsions.
4. What kind of bond do unsaturated vegetable oils contain?

Key Words   Emulsion • Emulsifier • Hydrophilic • Hydrophobic

C1 Changes in the Earth and its Atmosphere

Structure of the Earth

The **Earth** is nearly spherical. It has a layered structure that consists of…

- a thin **crust**
- a **mantle**
- a **core** (made of nickel and iron).

Rocks at the Earth's surface are continually being broken up, reformed and changed in an ongoing cycle of events, known as the **rock cycle**. The changes take a very long time.

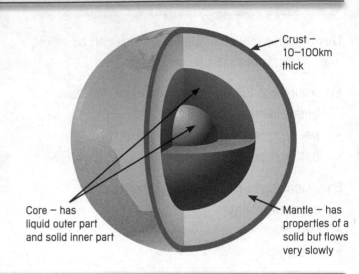

Crust – 10–100km thick

Core – has liquid outer part and solid inner part

Mantle – has properties of a solid but flows very slowly

Tectonic Theory

At one time, scientists believed that features on the Earth's surface, e.g. mountain ranges, were caused by shrinkage of the crust when the Earth cooled down following its formation.

But as scientists have found out more about the Earth this **theory** has been rejected.

Evidence showed scientists that the east coast of South America and the west coast of Africa have…

- **closely matching coastlines**
- **similar patterns of rocks**, which contain **fossils** of the same plants and animals, e.g. the Mesosaurus.

This evidence led Alfred Wegener to propose that South America and Africa had at one time been part of a single land mass. He proposed that the movement of the crust was responsible for the separation of the land, i.e. **continental drift**. This is **tectonic theory**.

Unfortunately, Wegener couldn't explain **how** the crust moved. It took more than 50 years for scientists to discover this.

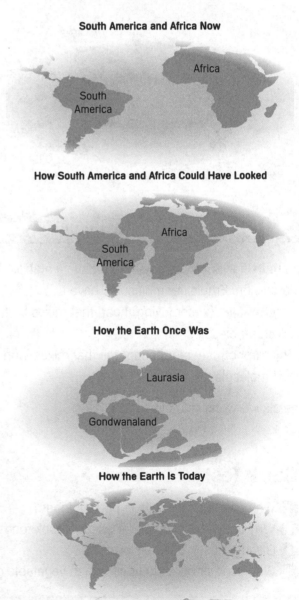

South America and Africa Now

Africa

South America

How South America and Africa Could Have Looked

Africa

South America

How the Earth Once Was

Laurasia

Gondwanaland

How the Earth Is Today

Key Words   **Theory • Evidence • Fossil**

Tectonic Plates

The Earth's lithosphere (the crust and the upper part of the mantle) is 'cracked' into **tectonic plates**.

Intense heat, released by radioactive decay deep in the Earth, creates convection currents in the mantle. These currents cause the tectonic plates to move apart very slowly, a few centimetres per year.

In convection in a gas or a liquid, the matter rises as it is heated. As it gets further away from the heat source, it cools and sinks. The same happens in the Earth:

1 Hot molten rock rises to the surface, creating new crust.

2 The older, cooler crust, then sinks down where the **convection current** starts to fall.

3 The land masses on these plates move slowly.

The movements are usually small and gradual. But sometimes they can be sudden and disastrous. **Earthquakes** and **volcanic eruptions** are common occurrences at plate boundaries. They are hard to predict.

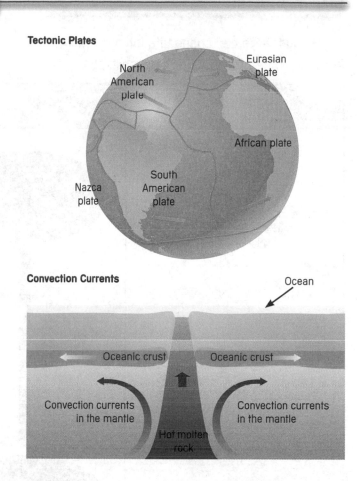

Tectonic Plates

The Origin of Life

There are many theories as to how life on Earth was formed because there is little direct evidence and therefore a number of assumptions have to be made.

(HT) Two scientists, **Miller** and **Urey**, tried to test one possible theory for how life on Earth began. They mixed together the chemicals thought to be present in the Earth's early atmosphere – water, methane and ammonia. The mixture was heated and sparks (electrical discharges) to represent ultraviolet radiation from the Sun were passed through it. The mixture was cooled and the cycle repeated many times.

After many cycles the mixture contained simple organic molecules, like amino acids, that are the building blocks of living organisms. This is known as the **primordial soup theory**, which is one (of many) that offers an explanation for how life began.

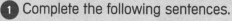

Quick Test

1 Complete the following sentences.

a) The Earth consists of a core, _____ and _____ .

b) Convection currents within the Earth's mantle cause the crust's plates to _____ slowly.

c) Earthquakes and/or volcanoes can occur at the boundaries between _____ _____ .

The Atmosphere

The **atmosphere** has changed a lot since the formation of the Earth 4.6 billion years ago.

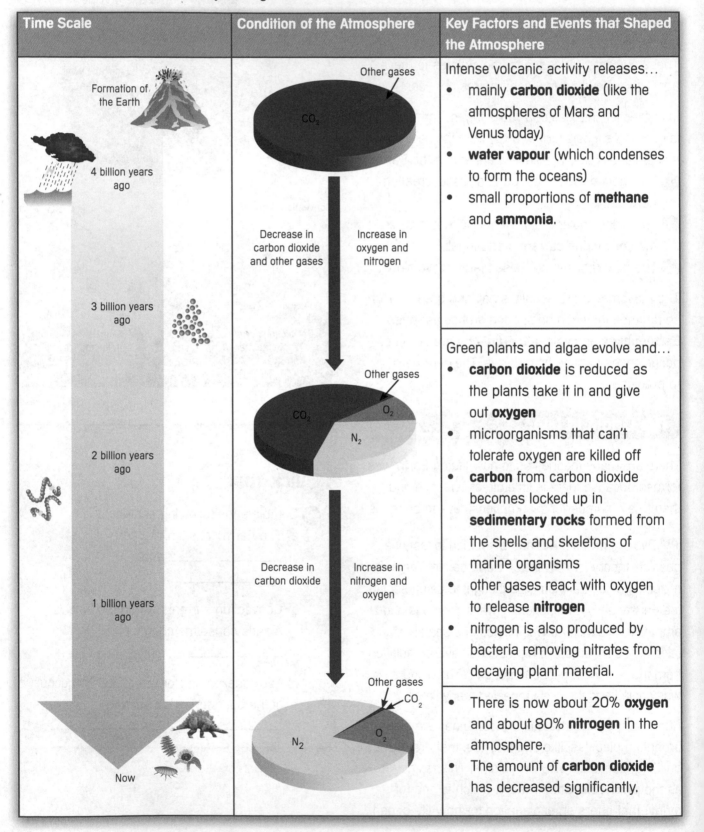

Time Scale	Condition of the Atmosphere	Key Factors and Events that Shaped the Atmosphere
Formation of the Earth / 4 billion years ago / 3 billion years ago / 2 billion years ago / 1 billion years ago / Now	*Other gases* / CO_2 — Decrease in carbon dioxide and other gases / Increase in oxygen and nitrogen	Intense volcanic activity releases… • mainly **carbon dioxide** (like the atmospheres of Mars and Venus today) • **water vapour** (which condenses to form the oceans) • small proportions of **methane** and **ammonia**.
	Other gases / CO_2 / O_2 / N_2 — Decrease in carbon dioxide / Increase in nitrogen and oxygen	Green plants and algae evolve and… • **carbon dioxide** is reduced as the plants take it in and give out **oxygen** • microorganisms that can't tolerate oxygen are killed off • **carbon** from carbon dioxide becomes locked up in **sedimentary rocks** formed from the shells and skeletons of marine organisms • other gases react with oxygen to release **nitrogen** • nitrogen is also produced by bacteria removing nitrates from decaying plant material.
	Other gases / CO_2 / O_2 / N_2	• There is now about 20% **oxygen** and about 80% **nitrogen** in the atmosphere. • The Amount of **carbon dioxide** has decreased significantly.

Composition of the Atmosphere

The proportions of gases in the atmosphere have been more or less the same for about 200 million years. The proportions are shown in the pie chart. **Water vapour** may also be present in varying quantities (0–3%).

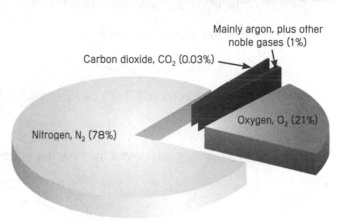

Mainly argon, plus other noble gases (1%)

Carbon dioxide, CO_2 (0.03%)

Oxygen, O_2 (21%)

Nitrogen, N_2 (78%)

Changes in the Atmosphere

The level of carbon dioxide in the atmosphere today is increasing due to...

- **volcanic activity** – geological activity moves carbonate rocks deep into the Earth and during volcanic activity they may release carbon dioxide back into the atmosphere
- **the burning of fossil fuels** – burning carbon, which has been locked up in **fossil fuels** for millions of years, releases carbon dioxide into the atmosphere.

The level of carbon dioxide in the atmosphere is reduced by the reaction between carbon dioxide and sea water. This reaction produces...

- insoluble carbonates that are deposited as sediment
- soluble hydrogencarbonates.

But too much carbon dioxide dissolving in the oceans can harm marine life.

The carbonates form some of the **sedimentary rocks** in the Earth's crust.

HT Fractional Distillation of Liquid Air

Air is a mixture of gases with different boiling points. The different gases can be collected by cooling air to a liquid and then heating it.

These gases are used as raw materials in a variety of industrial processes.

Nitrogen   Oxygen   Carbon Dioxide

Quick Test

1. **a)** Which two gases is the atmosphere mainly made up of?
   b) Name one other gas present in the atmosphere.
2. How is oxygen thought to have been released into the atmosphere?
3. What are the two main ways that carbon dioxide is absorbed from the Atmosphere?

Key Words   **Fossil fuel**

C1 Exam Practice Questions

1 This question is about sodium and oxygen. Sodium is a metal and oxygen is a non-metal. An atom of sodium contains 11 protons and an atom of oxygen contains 8 protons.

a) What charge do protons have?

.... **(1 mark)**

b) How many electrons will there be in an atom of sodium?

.... **(1 mark)**

c) Complete the diagram below to show the electronic structure of an atom of sodium.   **(1 mark)**

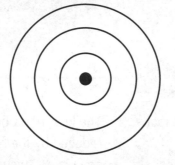

d) When sodium and oxygen react together sodium oxide is formed, which has the chemical formula Na_2O.

    i) Complete the following sentence.

       Na_2O contains a total of .... atoms.   **(1 mark)**

HT **ii)** Balance the equation below showing the formation of sodium oxide from sodium and oxygen.

$$Na \ + \ O_2 \ \longrightarrow \ Na_2O$$ **(1 mark)**

2 This question concerns some of the reactions of limestone. Limestone contains calcium carbonate ($CaCO_3$).

Limestone can be decomposed upon heating to form calcium oxide and carbon dioxide. Calcium oxide reacts with water to form calcium hydroxide. Calcium hydroxide solution is commonly known as limewater.

a) How is limestone obtained from the earth?

.... **(1 mark)**

b) Name another metal carbonate that decomposes upon heating in the same way as limestone.

.... **(1 mark)**

c) What would you observe if carbon dioxide gas is bubbled through limewater?

.... **(1 mark)**

3 Copper is a very useful metal that is mined and extracted from ores. Some ores are copper-rich. Other low-grade ores contain less copper. Copper can be used for electrical wiring.

a) Explain why copper is a useful material for use in electrical wiring.

.... **(1 mark)**

b) How is copper extracted from copper-rich ores?

.... **(1 mark)**

c) How is this copper then purified?

.... **(1 mark)**

d) Name one method used to extract copper from low-grade ores.

.... **(1 mark)**

4 Crude oil is a mixture of a large number of compounds. Many of these compounds are hydrocarbons. The combustion of hydrocarbon fuels releases heat but also many pollutant gases.

a) What do you understand by the term **mixture**?

.... **(1 mark)**

b) Which two elements do hydrocarbon compounds contain?

.... **(2 marks)**

c) Name a gas that may be released into the atmosphere when a hydrocarbon fuel burns.

.... **(1 mark)**

d) What name is given to the process of converting long-chain hydrocarbons into smaller, more useful molecules?

.... **(1 mark)**

5 Consider the two molecules (labelled as X and Y) shown below:

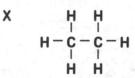
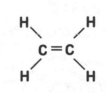

Which molecule, X or Y, is...

a) Saturated ....

b) An alkene ....

c) Used to make polymers ....

d) Able to decolourise bromine water ....

(4 marks)

C2 The Structure of Substances

Ionic Compounds

Ionic compounds are giant structures of **ions**. They are held together by **strong forces** of attraction (electrostatic forces) between **oppositely charged ions**, that act in **all directions**. This type of bonding is called **ionic bonding**.

Ionic compounds...

- have **high melting** and **boiling points**
- **conduct electricity** when molten or in solution because the charged ions are free to move about and carry the current.

+ Positively charged ion   - Negatively charged ion

The Ionic Bond

An **ionic bond** occurs between a **metal** and a **non-metal**. It involves a **transfer** of **electrons** from one atom to the other.

This forms electrically charged **ions**, each of which has a complete outer energy level.

Ions have the electronic structure of a noble gas.

- Atoms that **lose electrons** become **positively** charged ions.
- Atoms that **gain electrons** become **negatively** charged ions.

Example 1
Sodium (Na) and chlorine (Cl) bond ionically to form sodium chloride, NaCl.

1. The sodium atom has one electron in its outer shell.
2. The electron is transferred to the chlorine atom.
3. Both atoms now have eight electrons in their outer shell.
4. The atoms become ions, Na$^+$ and Cl$^-$.
5. The compound formed is sodium chloride, NaCl.

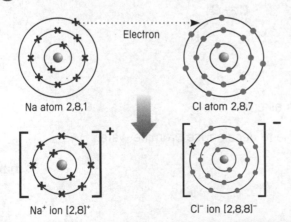

Na atom 2,8,1   Cl atom 2,8,7

Na$^+$ ion [2,8]$^+$ Cl$^-$ ion [2,8,8]$^-$

Example 2
Magnesium (Mg) and oxygen (O) bond ionically to form magnesium oxide, MgO.

1. The magnesium atom has two electrons in its outer shell.
2. These two electrons are transferred to the oxygen atom.
3. Both atoms now have eight electrons in their outer shell.
4. The atoms become ions, Mg^{2+} and O^{2-}.
5. The compound formed is magnesium oxide, MgO.

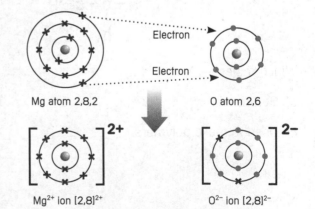

Mg atom 2,8,2   O atom 2,6

Mg^{2+} ion [2,8]$^{2+}$ O^{2-} ion [2,8]$^{2-}$

Ionic compound • Ion • Ionic bond • Electron

Alkali Metals and Halogens

The Alkali Metals (Group 1)

The alkali metals...
- have one electron in their outermost shell
- react with **non-metal elements** to form **ionic compounds** where the metal ion has a single **positive** charge.

Lithium Atom
2,1

Sodium Atom
2,8,1

Potassium Atom
2,8,8,1

The Halogens (Group 7)

The halogens...
- have seven electrons in their outermost shell
- react with **alkali metals** to form **ionic compounds** where the halide ions have a single **negative** charge.

Fluorine Atom
2,7

Chlorine Atom
2,8,7

Bromine Atom
2,8,8,7

Mixtures and Compounds

A **mixture** consists of two or more elements or **compounds** that are **not chemically combined**. The properties of the substances remain unchanged and specific to that substance.

Compounds are substances in which the atoms of two or more elements **are chemically combined** (not just mixed together).

Atoms can form chemical bonds by...
- **sharing electrons (covalent bonds)**
- **gaining** or **losing electrons (ionic bonds)**.

When atoms form **chemical bonds**, the arrangement of the **outer shell** of electrons **changes**. This results in each atom getting a **complete outer shell** of electrons. For most atoms this is eight electrons, but for helium it is only two.

Simple Molecular Compounds

Substances that consist of **simple molecules** are gases, liquids and solids that have relatively **low melting** and **boiling points**. The molecules have no overall electrical charge, so they can't conduct electricity.

HT Simple molecular compounds have low melting and boiling points because they have weak intermolecular forces (forces between their molecules).

Strong covalent bond within the molecule

Weak forces of attraction between molecules

Quick Test

1 What is a compound?
2 When atoms share a pair of electrons, what type of bond is formed?
3 a) Fill in the missing words:
   Simple covalent substances have relatively low _____ _____ and they have weak _____ forces between the molecules.
   b) When molten or dissolved in water, ionic compounds conduct _____ because the ions are free to _____ .

C2 The Structure of Substances

The Covalent Bond

A **covalent bond** occurs between **non-metal atoms**. It is a strong bond formed when **pairs of electrons are shared**.

Some covalently bonded substances have **simple structures**, e.g. H_2, Cl_2, O_2, HCl, H_2O and CH_4.

Others have **giant covalent structures**, called **macromolecules**, e.g. diamond and silicon dioxide.

Atoms that share electrons usually have **low melting** and **boiling points**. This is because they often form molecules in which there are…

- **strong covalent bonds** between the **atoms**
- **weak forces of attraction** between the **molecules**.

These forces are very weak compared to the strength of covalent bonds.

Example

1. A chlorine atom has seven electrons in its outer shell.
2. In order to bond with another chlorine atom, an electron from each atom is shared.
3. This gives each chlorine atom eight electrons in the outer shell.
4. Each atom now has a complete outer shell.

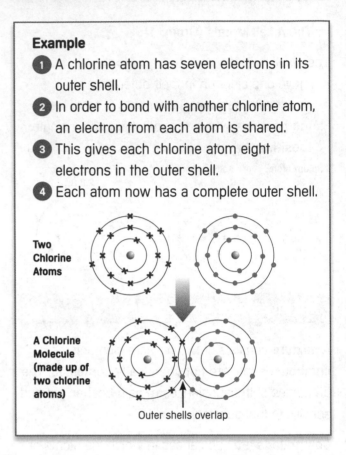

Two Chlorine Atoms

A Chlorine Molecule (made up of two chlorine atoms)

Outer shells overlap

Covalent Bonding

There are three different methods for representing the covalent bonds in each molecule. You need to be familiar with the following examples, and know how to use the different methods.

The two most common forms are shown in the table below.

The third form is shown here for an ammonia molecule. But, unless specifically asked for, you should use the other two methods.

Molecule	Water H_2O	Chlorine Cl_2	Hydrogen H_2	Hydrogen chloride, HCl	Methane CH_4	Oxygen O_2
Method 1	H O H	Cl Cl	H H	H Cl	H C H (with H top and bottom)	O O
Method 2	H–O–H	Cl – Cl	H – H	H – Cl	H–C–H (with H top and bottom)	O=O (a double bond)

Giant Covalent Structures

All the atoms in giant covalent structures are linked by **strong covalent bonds**. This means they have very **high melting points**.

Diamond is a form of carbon that has a **giant, rigid covalent structure** (lattice). Each carbon atom forms **four covalent bonds** with other carbon atoms. Diamond has a **large number of covalent bonds** so it's a very **hard substance**.

Graphite is a form of carbon that also has a giant covalent structure. However, in graphite, each carbon atom forms **three covalent bonds** with other carbon atoms in a layered structure. Graphite has layers that can slide past each other, making it soft and slippery.

HT The layers in graphite are held together by weak intermolecular forces. In graphite, one electron from each carbon atom is **delocalised**. These delocalised electrons allow graphite to **conduct heat** and **electricity**.

Silicon dioxide (or silica, SiO_2) has a lattice structure similar to diamond. Each **oxygen** atom is joined to **two silicon atoms**, and each **silicon** atom is joined to **four oxygen atoms**.

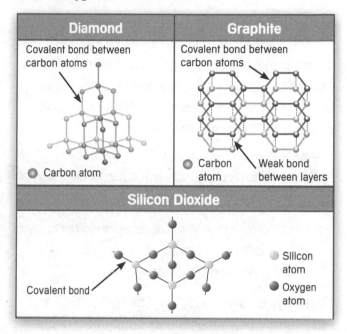

HT Fullerenes

Carbon can also form molecules known as **fullerenes**, which have different numbers of carbon atoms.

The structure of fullerenes is based on hexagon rings of carbon atoms.

Fullerenes can be used to deliver drugs in the body, in lubricants, as catalysts, and in **nanotubes** for reinforcing materials, e.g. in tennis rackets.

Metals

The layers of atoms in metals are able to slide over each other. This means that metals can be **bent and shaped**.

HT Metals have a giant structure in which electrons in the highest energy level can be delocalised. This produces a regular arrangement (lattice) of **positive ions** that are held together by electrons using electrostatic attraction.

These delocalised electrons can move around freely. This allows metals to **conduct heat and electricity**.

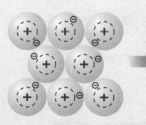
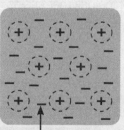

Delocalised electron

C2 The Structure of Substances

Alloys

An **alloy** is a mixture that contains a **metal** and at least one other **element**.

The added element disturbs the regular arrangement of the metal atoms so the layers don't slide over each other so easily. So, **alloys** are usually stronger and harder than pure metal.

Shape Memory Alloys

Smart alloys belong to a group of materials that are being developed to meet the demands of modern engineering and manufacturing. These materials respond to changes in their environment.

Smart alloys **remember their shape**. They can be deformed but will return to their original shape, for example, flexible spectacle frames. Nitinol (used in dental braces) is an example of a smart alloy.

Properties of Polymers

The properties of polymers depend on...
- what they're made from (i.e. what **monomer** is used)
- the conditions (i.e. temperature and catalyst) under which they're made.

For example, low density poly(ethene) (LDPE) and high density poly(ethene) (HDPE) are both made from the monomer ethene. But the polymers have different properties because different catalysts and reaction conditions are used to make them.

LDPE is used to make carrier bags and HDPE is used to make plastic bottles.

Thermo-softening and Thermo-setting Polymers

Thermo-softening polymers consist of individual polymer chains that are tangled together (like spaghetti).

(HT) There are weak intermolecular forces between all of the polymer chains in a thermo-softening polymer, which helps to explain its properties.

Thermo-setting polymers consist of polymer chains that are joined together by cross-links between them. Thermo-setting polymers don't melt when they're heated.

Thermo-setting Polymer

Cross-links

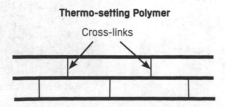

Thermo-softening Polymer (no cross-links)

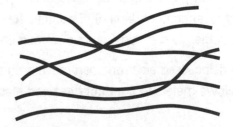

Nanoparticles and Nanostructures

Nanoscience is the study of very small structures. The structures are 1–100 nanometres in size, roughly in the order of a few hundred atoms.

One **nanometre** is 0.000 000 001m (one billionth of a metre) and is written as 1nm or 1×10^{-9}m.

Nanoparticles are tiny, tiny particles that can combine to form structures called **nanostructures**.

Nanostructures can be manipulated so materials can be developed that have new and specific properties.

The **properties** of **nanoparticles** are different to the properties of the **same materials in bulk**.

For example…
- nanoparticles are more sensitive to light, heat and magnetism
- nanoparticles have a high surface area in relation to their volume.

Research into nanoparticles may lead to the development of new…
- computers
- catalysts
- coatings
- highly selective sensors
- stronger and lighter construction materials
- cosmetics, e.g. suntan creams and deodorants.

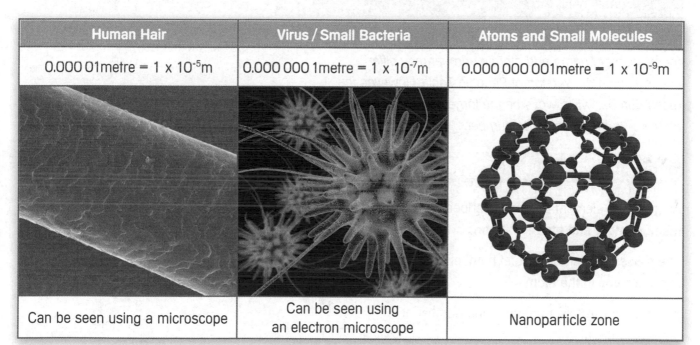

Human Hair	Virus / Small Bacteria	Atoms and Small Molecules
0.000 01metre = 1×10^{-5}m	0.000 000 1metre = 1×10^{-7}m	0.000 000 001metre = 1×10^{-9}m
Can be seen using a microscope	Can be seen using an electron microscope	Nanoparticle zone

Quick Test

1. Why are metals able to be bent and shaped?
2. What is the main structural difference between thermo-softening and thermo-setting polymers?
3. Give two potential uses of nanoparticles.

Key Words Nanoscience

The Periodic Table

1	2											3	4	5	6	7	8 or 0
						1 H hydrogen 1											4 He helium 2
7 Li lithium 3	9 Be beryllium 4											11 B boron 5	12 C carbon 6	14 N nitrogen 7	16 O oxygen 8	19 F fluorine 9	20 Ne neon 10
23 Na sodium 11	24 Mg magnesium 12											27 Al aluminium 13	28 Si silicon 14	31 P phosphorus 15	32 S sulfur 16	35.5 Cl chlorine 17	40 Ar argon 18
39 K potassium 19	40 Ca calcium 20	45 Sc scandium 21	48 Ti titanium 22	51 V vanadium 23	52 Cr chromium 24	55 Mn manganese 25	56 Fe iron 26	59 Co cobalt 27	59 Ni nickel 28	63.5 Cu copper 29	65 Zn zinc 30	70 Ga gallium 31	73 Ge germanium 32	75 As arsenic 33	79 Se selenium 34	80 Br bromine 35	84 Kr krypton 36
85 Rb rubidium 37	88 Sr strontium 38	89 Y yttrium 39	91 Zr zirconium 40	93 Nb niobium 41	96 Mo molybdenum 42	[98] Tc technetium 43	101 Ru ruthenium 44	103 Rh rhodium 45	106 Pd palladium 46	108 Ag silver 47	112 Cd cadmium 48	115 In indium 49	119 Sn tin 50	122 Sb antimony 51	128 Te tellurium 52	127 I iodine 53	131 Xe xenon 54
133 Cs caesium 55	137 Ba barium 56	139 La* lanthanum 57	178 Hf hafnium 72	181 Ta tantalum 73	184 W tungsten 74	186 Re rhenium 75	190 Os osmium 76	192 Ir iridium 77	195 Pt platinum 78	197 Au gold 79	201 Hg mercury 80	204 Tl thallium 81	207 Pb lead 82	209 Bi bismuth 83	[209] Po polonium 84	[210] At astatine 85	[222] Rn radon 86
[223] Fr francium 87	[226] Ra radium 88	[227] Ac* actinium 89	[261] Rf rutherfordium 104	[262] Db dubnium 105	[266] Sg seaborgium 106	[264] Bh bohrium 107	[277] Hs hassium 108	[268] Mt meitnerium 109	[271] Ds darmstadtium 110	[272] Rg roentgenium 111							

N.B. The exact position of the mass number/relative atomic mass, element name and atomic number may differ depending on the version of Periodic Table. However, the mass number will always be the larger number, and the atomic number the smaller number.

Mass Number and Atomic Number

Atoms of an element can be described using their **mass number** and **atomic number**.

The **mass number** is the total number of **protons** and **neutrons** in the atom.

The **atomic (proton) number** is the number of protons in the atom.

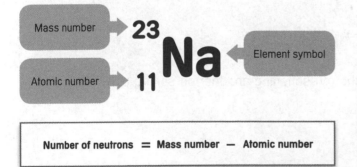

Mass number → **23** **Na** ← Element symbol

Atomic number → **11**

Number of neutrons = Mass number — Atomic number

The number of **protons** in an atom is **equal** to the number of **electrons**. So, an atom has **no overall charge**.

Examples

Hydrogen

$^{1}_{1}$H

1 proton
1 electron

Oxygen

$^{16}_{8}$O

8 protons
8 electrons

Although they have the same charge, protons and electrons have a **different mass**.

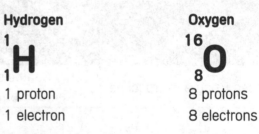

Atomic Particle	Relative Mass
Proton	1
Neutron	1
Electron	Very small (negligible)

Isotopes

All atoms of a **particular element** have the **same number** of protons. Atoms of **different elements** have **different numbers** of protons.

Isotopes are atoms of the **same element** that have **different numbers of neutrons**.

Isotopes have the **same atomic number** but a **different mass number**.

For example, chlorine has two isotopes.

$$^{35}_{17}\text{Cl}$$

17 protons
17 electrons
18 neutrons (35 − 17)

$$^{37}_{17}\text{Cl}$$

17 protons
17 electrons
20 neutrons (37 − 17)

Relative Atomic Mass, A_r

The **relative atomic mass, A_r,** of an element is found on the Periodic Table. It is the larger number shown for each element.

Relative atomic mass → $$^{16}_{8}\text{O}$$

(HT) The relative atomic mass, A_r, is the mass of a particular atom compared with a twelfth of the mass of a carbon atom (the ^{12}C isotope).

The A_r is an **average** value for all the **isotopes** of the element.

By looking at the Periodic Table, you can see that...
• carbon is 12 times heavier than hydrogen, but is only half as heavy as magnesium
• magnesium is three-quarters as heavy as sulfur
• sulfur is twice as heavy as oxygen, etc.

You can use this idea to calculate the **relative formula mass** of compounds.

Relative Formula Mass, M_r

The **relative formula mass, M_r,** of a compound is the relative atomic masses of all its elements added together.

To calculate M_r, you need to know...
• the formula of the compound
• the A_r of all the atoms involved.

Example 1
Calculate the M_r of water, H_2O.

Write the formula → **H_2O**

Substitute the A_rs → $(2 \times 1) + 16$

Calculate the M_r → $2 + 16 = 18$

Example 2
Calculate the M_r of potassium carbonate, K_2CO_3.

Write the formula → **K_2CO_3**

Substitute the A_rs → $(39 \times 2) + 12 + (16 \times 3)$

Calculate the M_r → $78 + 12 + 48 = 138$

Quick Test

1. What is the total number of protons and neutrons in an atom called?
2. What are atoms of the same element that have different numbers of neutrons called?

Calculating Percentage Mass

The mass of the compound is its **relative formula mass** in grams.

To calculate the **percentage mass** of an element in a compound, you need to know...
* the **formula** of the compound
* the **relative atomic mass** of all the atoms.

You can calculate the percentage mass by using this formula:

$$\frac{\text{Relative mass of element in the compound}}{\text{Relative formula mass of compound (M}_r\text{)}} \times 100$$

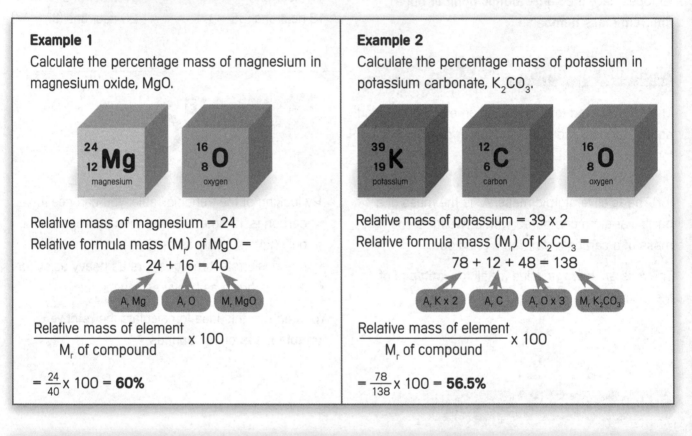

Example 1

Calculate the percentage mass of magnesium in magnesium oxide, MgO.

$^{24}_{12}$Mg magnesium $^{16}_{8}$O oxygen

Relative mass of magnesium = 24
Relative formula mass (M_r) of MgO =
24 + 16 = 40

A_r Mg A_r O M_r MgO

$$\frac{\text{Relative mass of element}}{M_r \text{ of compound}} \times 100$$

$$= \frac{24}{40} \times 100 = \textbf{60\%}$$

Example 2

Calculate the percentage mass of potassium in potassium carbonate, K_2CO_3.

$^{39}_{19}$K potassium $^{12}_{6}$C carbon $^{16}_{8}$O oxygen

Relative mass of potassium = 39 x 2
Relative formula mass (M_r) of K_2CO_3 =
78 + 12 + 48 = 138

A_r K x 2 A_r C A_r O x 3 M_r K_2CO_3

$$\frac{\text{Relative mass of element}}{M_r \text{ of compound}} \times 100$$

$$= \frac{78}{138} \times 100 = \textbf{56.5\%}$$

HT Empirical Formula of a Compound

The empirical formula of a compound is the **simplest whole number ratio** of each kind of atom in the compound.

Example

Find the simplest formula of an oxide of iron produced by reacting 1.12g of iron with 0.48g of oxygen (A_r Fe = 56, A_r O = 16).

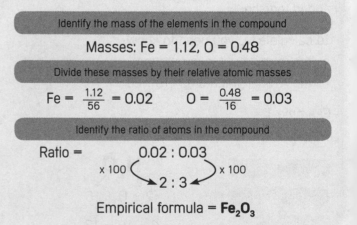

Identify the mass of the elements in the compound

Masses: Fe = 1.12, O = 0.48

Divide these masses by their relative atomic masses

Fe = $\frac{1.12}{56}$ = 0.02   O = $\frac{0.48}{16}$ = 0.03

Identify the ratio of atoms in the compound

Ratio =   0.02 : 0.03
x 100   2 : 3   x 100

Empirical formula = **Fe$_2$O$_3$**

The Mole

A **mole** (mol) is a measure of the **number of particles** (atoms or molecules) contained in a substance. One mole of a substance is its relative formula mass or A_r in grams.

One mole of **any substance** (element or compound) will always contain the **same number** of particles – six hundred thousand billion billion or 6×10^{23}. This is the **relative formula mass** of the substance.

If a substance is an **element**, the mass of one mole of the substance, called the molar mass (g/mol), is always **equal** to the **relative atomic mass** of the substance in grams. For example...

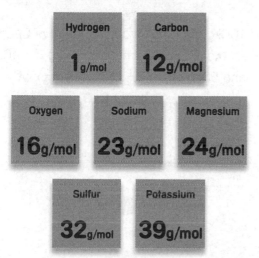

Hydrogen	Carbon
1g/mol	**12**g/mol

Oxygen	Sodium	Magnesium
16g/mol	**23**g/mol	**24**g/mol

Sulfur	Potassium
32g/mol	**39**g/mol

If a substance is a **compound**, the mass of one mole of the substance is always **equal** to the **relative formula mass** of the substance in grams. For example, one mole of sodium hydroxide (NaOH)...

A_r sodium + A_r hydrogen + A_r oxygen

= 23 + 1 + 16 = **40g**

You can calculate the number of moles in a substance using this formula:

$$\text{Number of moles of substance (mol)} = \frac{\text{Mass of substance (g)}}{\text{Mass of one mole (g/mol)}}$$

N.B. You need to remember this equation as it will not be given to you in the exam.

Example 1

Calculate the number of moles of carbon in 36g of the element.

$$\text{Number of moles of substance (mol)} = \frac{\text{Mass of substance (g)}}{\text{Mass of one mole (g/mol)}}$$

$$= \frac{36g}{12g/mol} \quad \leftarrow A_r \text{ carbon} = 12$$

$$= \textbf{3 moles}$$

Example 2

Calculate the number of moles of carbon dioxide in 33g of the gas.

$$\text{Number of moles of substance (mol)} = \frac{\text{Mass of substance (g)}}{\text{Mass of one mole (g/mol)}}$$

$$= \frac{33g}{44g/mol} \quad \leftarrow$$

A_r carbon dioxide = A_r carbon + 2 x A_r oxygen = 12 + (2 x 16) = 44

$$= \textbf{0.75 mole}$$

Example 3

Calculate the mass of four moles of sodium hydroxide.

$$\text{Mass of substance (g)} = \text{Number of moles of substance (mol)} \times \text{Mass of one mole (g/mol)}$$

$$= \text{4mol} \times \text{40g/mol}$$

$$= \textbf{160g}$$

N.B. If you are confident in your mathematical ability, you can also do these calculations using ratios.

Quick Test

1. Calculate the percentage by mass of nitrogen (N) in each of the following compounds:
   a) HNO_3 **b)** NH_4NO_3 **c)** $(NH_4)_2SO_4$
2. Calculate the number of moles of the following:
   a) 48g of magnesium   **b)** 12.4g of SO_2 gas
3. Calculate the mass of the following:
   a) Two moles of carbon monoxide (CO)
   b) Three moles of potassium carbonate (K_2CO_3)

HT Calculating the Mass of a Product

Example

Calculate how much calcium oxide can be produced from 50kg of calcium carbonate. (Relative atomic masses: Ca = 40, C = 12, O = 16).

1 Write down the equation.

2 Work out the M_r of each substance.

3 Check that the total mass of reactants equals the total mass of the products. If they are not the same, check your work.

4 The question only mentions calcium oxide and calcium carbonate, so you can now ignore the carbon dioxide. You just need the ratio of mass of reactant to mass of product.

5 Use the ratio to calculate how much calcium oxide can be produced.

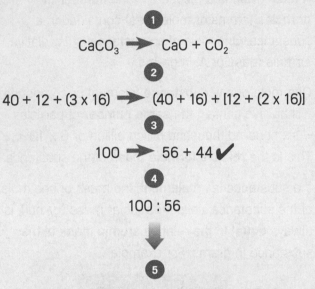

1
$$CaCO_3 \rightarrow CaO + CO_2$$

2
$$40 + 12 + (3 \times 16) \rightarrow (40 + 16) + [12 + (2 \times 16)]$$

3
$$100 \rightarrow 56 + 44 \checkmark$$

4
$$100 : 56$$

5

If 100kg of $CaCO_3$ produces 56kg of CaO, then 1kg of $CaCO_3$ produces $\frac{56}{100}$ kg of CaO, and 50kg of $CaCO_3$ produces $\frac{56}{100} \times 50$

= 28kg of CaO

HT Calculating the Mass of a Reactant

Example

Calculate how much aluminium oxide is needed to produce 540 tonnes of aluminium. (Relative atomic masses: Al = 27, O = 16).

1 Write down the equation.

2 Work out the M_r of each substance.

3 Check that the total mass of reactants equals the total mass of the products. If they are not the same, check your work.

4 The question only mentions aluminium oxide and aluminium, so you can now ignore the oxygen. You just need the ratio of mass of reactant to mass of product.

5 Use the ratio to calculate how much aluminium oxide is needed.

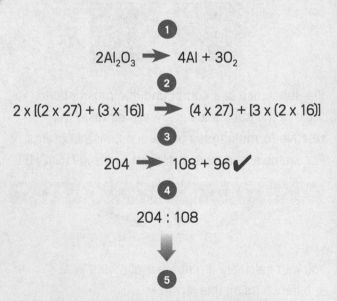

1
$$2Al_2O_3 \rightarrow 4Al + 3O_2$$

2
$$2 \times [(2 \times 27) + (3 \times 16)] \rightarrow (4 \times 27) + [3 \times (2 \times 16)]$$

3
$$204 \rightarrow 108 + 96 \checkmark$$

4
$$204 : 108$$

5

If 204 tonnes of Al_2O_3 produces 108 tonnes of Al, then $\frac{204}{108}$ tonnes is needed to produce 1 tonne of Al, and $\frac{204}{108} \times 540$ tonnes is needed to produce 540 tonnes of Al

= 1020 tonnes of Al_2O_3

Instrumental Methods

Standard laboratory equipment can be used to detect and **identify elements** and **compounds**. But **instrumental methods** that involve using highly **accurate** instruments to analyse and identify substances have been developed to perform this function.

These instruments give rapid results, are very sensitive and accurate, and can be used on small samples.

An example of a common instrumental method of analysis is Gas Chromatography linked to Mass Spectroscopy (GC-MS).

A GC-MS works by allowing different substances, carried by a gas, to travel through a column packed with solid material at different speeds so that they separate out. Each substance will produce a separate peak on an output known as a gas chromatograph. The number of peaks on this output shows the number of compounds present in the original sample.

The position of the peaks on the output graph indicates the retention time, i.e. the time taken to pass through the gas chromatograph.

If the output of the gas chromatography column is linked to a mass spectrometer then this can also be used to identify the substances leaving the column.

(HT) The mass spectrometer can give the relative molecular mass (M_r) of each substance separated in the column.

The molecular mass is given by the molecular ion peak on the spectrum.

Chromatography

Chemical analysis can be used to identify additives in food. **Chromatography** is used to identify artificial colours, by comparing them to known substances.

1. Samples of five known food colourings (A, B, C, D and E), and the unknown substance (X) are put on a 'start line' on a piece of paper.

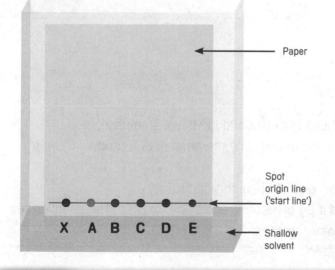

Paper

Spot origin line ('start line')

X A B C D E

Shallow solvent

2. The paper is dipped into a solvent. The solvent dissolves the samples and carries them up the paper.

3. Substance X can be identified by comparing the horizontal spots.

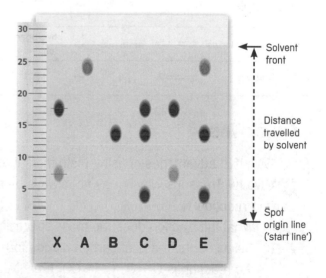

Solvent front

Distance travelled by solvent

Spot origin line ('start line')

X A B C D E

We can see that substance X is food colouring D

Yield

Atoms are **never lost or gained** in a chemical reaction. But, it isn't always possible to obtain the calculated amount of the product for several reasons:

- If the reaction is reversible, it might not go to completion.
- Some product could be lost when it's separated from the reaction mixture.
- Some of the reactants may react in different ways to the expected reaction.

The amount of product obtained is called the **yield**.

The **percentage yield** can be calculated by comparing...

- the actual yield obtained from a reaction
- the maximum theoretical yield.

(HT)

$$\text{Percentage yield} = \frac{\text{Yield from reaction}}{\text{Maximum theoretical yield}} \times 100$$

Example

50kg of calcium carbonate ($CaCO_3$) is expected to produce 28kg of calcium oxide (CaO).

A company heats 50kg of calcium carbonate in a kiln and obtains 22kg of calcium oxide.

Calculate the percentage yield.

Percentage yield = $\frac{22}{28}$ x 100

$$= \textbf{78.6\%}$$

Reversible Reactions

Some chemical reactions are **reversible**. In a **reversible reaction**, the **products** can **react** to produce the **original reactants**.

These reactions are represented as...

$$ A + B \rightleftharpoons C + D $$

This means that A and B can react to produce C and D, and C and D can also react to produce A and B.

For example...

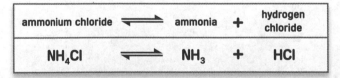

ammonium chloride $\rightleftharpoons$	ammonia	+	hydrogen chloride
NH_4Cl $\rightleftharpoons$	NH_3	+	HCl

Solid ammonium chloride decomposes when heated to produce ammonia and hydrogen chloride gas (both colourless).

Ammonia reacts with hydrogen chloride gas to produce clouds of white ammonium chloride powder.

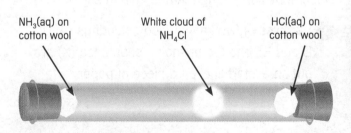

NH_3(aq) on cotton wool — White cloud of NH_4Cl — HCl(aq) on cotton wool

Quick Test

1. Give two advantages of using instrumental methods to detect and identify elements.
2. Give two reasons why it may not always be possible to calculate the amount of a product formed in a reaction.
3. **HT** What information does the mass spectrometer give about a compound?
4. **HT** Calculate the mass of carbon dioxide formed if 3g of carbon reacts with oxygen. The equation for the reaction is C + O_2 $\longrightarrow$ CO_2 (Relative atomic masses: C = 12, O = 16)

Rates of Reactions

Chemical reactions only occur when reacting particles **collide** with each other with **sufficient energy**.

The **minimum amount** of energy required to cause a reaction is called the **activation energy**.

There are four important factors that affect the rate of reaction:

- Temperature.
- Concentration.
- Surface area.
- Use of a catalyst.

Temperature

In a **cold** reaction mixture the particles move quite **slowly**. They collide less often, with less energy, so **fewer collisions** are successful.

In a **hot** reaction mixture the particles move more **quickly**. They collide more often, with greater energy, so **more collisions** are successful.

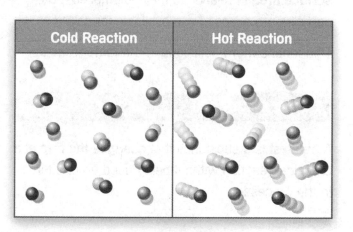

Cold Reaction	Hot Reaction

Concentration

In a **low concentration** reaction, the particles are **spread out**. They collide less often, so there are fewer successful collisions.

In a **high concentration** reaction, the particles are crowded **close together**. They collide more often, so there are more successful collisions.

Increasing the **pressure** of reacting gases also increases the frequency of collisions.

(HT) **Concentrations** of solutions are given in **moles per cubic decimetre (mol/dm³)**.

Equal volumes of solutions of the same molar concentration contain the same number of moles of solute, i.e. the same number of particles.

Equal volumes of gases at the same temperature and pressure contain the **same number** of particles.

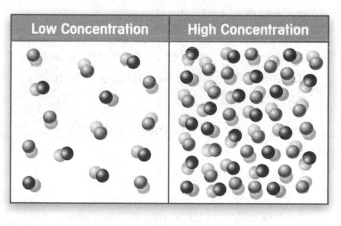

Low Concentration	High Concentration

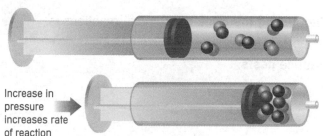

Increase in pressure increases rate of reaction

C2 Rates of Reaction

Surface Area

Large pieces of a solid reactant have a **small surface area** in relation to their volume.

Fewer particles are exposed and available for collisions. This means **fewer collisions** and a **slower reaction**.

Small pieces of a solid reactant have a **large surface area** in relation to their volume, so more particles are exposed and available for collisions.

This means **more collisions** and a **faster reaction**.

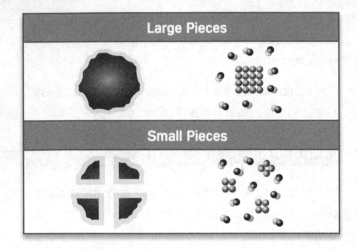

Large Pieces

Small Pieces

Using a Catalyst

A **catalyst** is a substance that **changes the rate** of a chemical reaction without being used up or altered in the process.

A catalyst…

* reduces the amount of energy needed for a successful collision
* makes more collisions successful
* speeds up the reaction
* provides a surface for the molecules to attach to, which increases their chances of bumping into each other.

Different reactions need different catalysts. For example…

* the cracking of hydrocarbons uses broken pottery
* the manufacture of ammonia uses iron.

Increasing the rates of chemical reactions is important in industry because it helps to **reduce costs**.

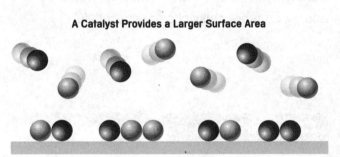

A Catalyst Provides a Larger Surface Area

Catalyst

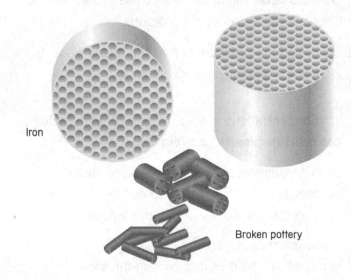

Catalysts Used in Industrial Reactions

Iron

Broken pottery

Analysing the Rate of Reaction

| Rate of Reaction | = | Amount of reactant used OR product formed / Time |

$$\text{Rate of Reaction} = \frac{\text{Amount of reactant used OR product formed}}{\text{Time}}$$

The rate of a chemical reaction can be found in two ways:

1 **Measuring the amount of reactants used.**
If one of the products is a gas, you could weigh the reaction mixture before and after the reaction takes place. The mass of the mixture will decrease.

2 **Measuring the amount of products formed.**
You could use a gas syringe to measure the total volume of gas produced at timed intervals.

Plotting Reaction Rates

Graphs can be plotted to show the progress of a chemical reaction. There are three things to remember:

1 The steeper the line, the faster the reaction.

2 When one of the reactants is used up, the reaction stops (the line becomes horizontal).

3 The same amount of product is formed from the same amount of reactants, irrespective of rate.

The graph shows us that reaction **A** is faster than reaction **B**. This could be due to several factors, including:

- The surface area of the solid reactants in **A** is greater than in **B**.
- The temperature of reaction **A** is greater than reaction **B**.
- The concentration of the solution in **A** is greater than in **B**.
- A catalyst is used in reaction **A** but not in reaction **B**.

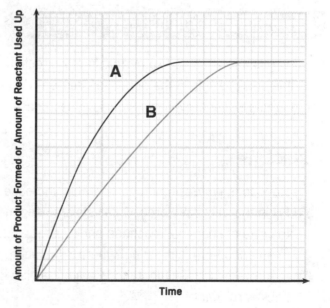

C2 Exothermic and Endothermic Reactions

Chemical Reactions

When chemical reactions occur, **energy** is transferred **to** or **from** the **surroundings**.

Many chemical reactions are, therefore, accompanied by a **temperature change**.

Exothermic Reactions

Exothermic reactions are accompanied by a **temperature rise**. They transfer heat energy to the surroundings, i.e. they **give out** heat.

Common examples of exothermic reactions include…
- neutralising alkalis with acids
- oxidation
- combustion
- self-heating can (e.g. for coffee)
- handwarmers.

methane (natural gas) + oxygen → carbon dioxide + water + heat energy
$CH_4 + 2O_2 \rightarrow CO_2 + 2H_2O$

carbon + oxygen → carbon dioxide + heat energy
$C + O_2 \rightarrow CO_2$

Endothermic Reactions

Endothermic reactions are accompanied by a **fall in temperature**. Heat energy is transferred from the surroundings, i.e. they **take in** heat.

Thermal decomposition and dissolving ammonium nitrate crystals in water are examples of endothermic reactions.

ammonium nitrate + water + heat energy → ammonium nitrate solution
$NH_4NO_3 + H_2O \rightarrow NH_4NO_3$

Some sports injury packs are based on endothermic reactions.

Reversible Reactions

If a reversible reaction is exothermic in one direction then it is endothermic in the opposite direction. The same amount of energy is transferred in each case.

Example

hydrated copper sulfate (blue)	⇌ endothermic / exothermic	anhydrous copper sulfate (white) + water

Quick Test

1. How does increasing the temperature increase the rate of a chemical reaction?
2. Fill in the missing words: Catalysts are used in industrial reactions because they _____ the rate of chemical reactions and _____ costs.
3. In an endothermic reaction, where is the energy absorbed from?
4. In a reversible reaction if the forward reaction is exothermic what can you conclude about the backward reaction?

State Symbols

State symbols are used in equations. The symbols are (s) **solid**, (l) **liquid**, (g) **gas** and (aq) **aqueous**.

An **aqueous solution** is produced when a substance is **dissolved in water**.

Soluble Salts from Metals

Metals react with dilute acid to form a **metal salt** and **hydrogen**.

Salt is a word used to describe any metal compound made from a reaction between a metal and an acid.

metal + acid ⟶ salt + hydrogen

Some metals react with acid more vigorously than others:

- Silver – no reaction.
- Zinc – fairly reasonable reaction.
- Magnesium – vigorous reaction.
- Potassium – very violent and dangerous reaction.

Soluble Salts from Insoluble Bases

Bases are the oxides and hydroxides of metals. **Soluble** bases are called **alkalis**.

The oxides and hydroxides of transition metals are **insoluble**. Their salts are prepared in the following way:

1. The metal oxide or hydroxide is added to an acid until no more will react.
2. The excess metal oxide or hydroxide is then filtered, leaving a solution of the salt.
3. The salt solution is then evaporated.

This reaction can be written generally as follows:

acid + base ⟶ neutral salt solution + water

Example

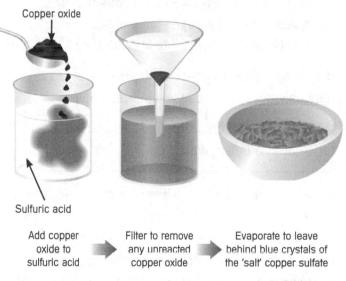

Copper oxide

Sulfuric acid

Add copper oxide to sulfuric acid ➡ Filter to remove any unreacted copper oxide ➡ Evaporate to leave behind blue crystals of the 'salt' copper sulfate

Salts of Alkali Metals

Compounds of alkali metals, called **salts**, can be made by reacting solutions of their hydroxides (which are alkaline) with a particular acid. This neutralisation reaction can be represented as follows:

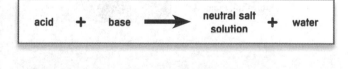

acid + alkaline hydroxide solution ⟶ neutral salt solution + water

The salt produced depends on the metal in the alkali and the acid used.

	Hydrochloric Acid	Sulfuric Acid	Nitric Acid
Metal Hydroxide	Metal chloride	Metal sulfate	Metal nitrate

C2 Acids, Bases and Salts

Insoluble Salts

Insoluble salts can be made by mixing appropriate solutions of ions so that a **precipitate** (solid substance) is formed.

Precipitation can be used to remove unwanted ions from a solution, e.g. softening hard water. The calcium (or magnesium) ions are precipitated out as insoluble calcium (or magnesium) carbonate.

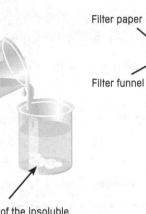

1. Two solutions of soluble substance are mixed together in a beaker

2. The precipitate is filtered off, rinsed and dried

Filter paper

Filter funnel

Precipitate

Precipitate of the insoluble salt is formed

Filtrate

Neutralisation

Acids and **alkalis** are **chemical opposites**:
- **Acids** contain **hydrogen ions**, H^+(aq).
- **Alkalis** contain **hydroxide ions**, OH^-(aq).

If they are added together in the correct amounts they can 'neutralise' (cancel out) each other.

When an acid reacts with an alkali, the **ions** react together to produce **water** (pH 7).

$$H^+{(aq)} + OH^-{(aq)} \longrightarrow H_2O_{(l)}$$

This type of reaction is called **neutralisation** because the solution that remains has a pH of 7, showing it is neutral. For example, hydrochloric acid and potassium hydroxide can be neutralised.

hydrochloric acid	+	potassium hydroxide	→	potassium chloride	+	water

$$HCl_{(aq)} + KOH_{(aq)} \longrightarrow KCl_{(s)} + H_2O_{(l)}$$

Neutralising HCl and KOH

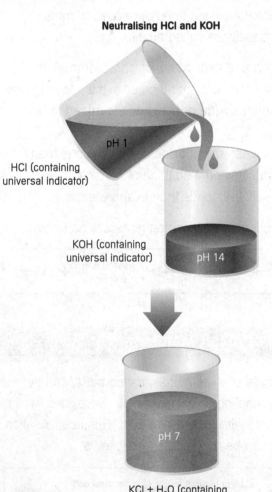

HCl (containing universal indicator)

pH 1

KOH (containing universal indicator)

pH 14

pH 7

KCl + H₂O (containing universal indicator)

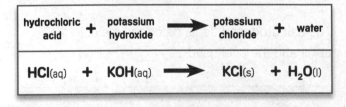

Key Words   Precipitate • Precipitation • Acid • Alkali • Neutralisation

Neutralising Ammonia

Ammonia is an **alkaline gas** that dissolves in water to make an **alkaline solution**.

It's mainly used in the production of fertilisers to increase the nitrogen content of the soil.

Ammonia neutralises nitric acid to produce **ammonium nitrate**. The aqueous ammonium nitrate is then evaporated to dryness.

| ammonia | + | nitric acid | ⟶ | ammonium nitrate |

$$NH_3(aq) + HNO_3(aq) \longrightarrow NH_4NO_3(aq)$$

Ammonium nitrate, a fertiliser rich in nitrogen, is also known as 'nitram' (nitrate of ammonia). **Nitrogen-based fertilisers** are important chemicals because they increase the yield of crops.

But nitrates can create problems if they find their way into streams, rivers or groundwater. Nitrates can…
- upset the natural balance of water
- contaminate our drinking water.

Ammonium hydroxide can be neutralised with acids to produce ammonium salts.

	Hydrochloric Acid	Sulfuric Acid	Nitric Acid
Ammonium Hydroxide	Ammonium chloride and water	Ammonium sulfate and water	Ammonium nitrate and water

Indicators and pH Scale

Indicators are dyes that change colour depending on whether they are in **acidic** or **alkaline** solutions:
- **Litmus** is an indicator that changes colour from red to blue or vice versa.
- **Universal indicator** is a mixture of dyes that show a **range** of colours to indicate **how** acidic or alkaline a substance is.

The **pH scale** is a measure of the **acidity or alkalinity** of an **aqueous solution**, across a **14-point scale**. When substances dissolve in water, they dissociate into their individual ions:
- Hydroxide ions, OH⁻(aq), make solutions alkaline.
- Hydrogen ions, H⁺(aq), make solutions acidic.

Very acidic					Neutral				Very alkaline				
1	2	3	4	5	6	7	8	9	10	11	12	13	14

Quick Test

1. By what kind of reaction are insoluble salts formed?
2. Fill in the missing words:
   a) Soluble hydroxides are called _____ .
   b) Ammonium salts are used as _____ .
   c) In neutralisation reactions H⁺ ions react with OH⁻ to form _____ .

C2 Electrolysis

Electrolysis

Electrolysis is the **breaking down** of a compound containing **ions** into its **elements** using an **electrical current**. The substance being broken down is called the **electrolyte**.

Ionic substances are chemical compounds that allow an **electric current** to flow through them when they are...

- molten
- dissolved in water.

These compounds contain **negative** and **positive** ions. During electrolysis...

- **negatively charged ions** move to the **positive electrode**
- **positively charged ions** move to the **negative electrode**.

When this happens, simpler substances are released at the two electrodes.

This moving of electrons forms electrically **neutral** atoms or molecules that are then released.

If there is a **mixture of ions** in the solution, the products formed depend on the reactivity of the elements involved.

For example, in the electrolysis of **copper chloride solution**, the simple substances released are...

- copper at the negative electrode
- chlorine gas at the positive electrode.

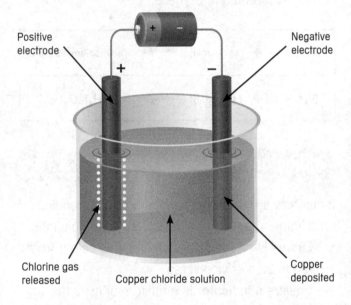

Positive electrode

Negative electrode

Chlorine gas released

Copper chloride solution

Copper deposited

Redox Reactions

Reduction is when **positively** charged ions **gain** electrons at the **negative** electrode.

Oxidation is when **negatively** charged ions **lose** electrons at the **positive** electrode.

A **redox reaction** is a chemical reaction where **both reduction and oxidation** occur.

You can remember this by thinking of the word **oilrig**:

- **O**xidation **I**s **L**oss of electrons (**OIL**).
- **R**eduction **I**s **G**ain of electrons (**RIG**).

Electroplating

Electrolysis can be used to electroplate objects with metals such as copper or silver.

Extraction of Aluminium

Aluminium is obtained by the electrolysis of aluminium oxide that has been mixed with **cryolite**. The cryolite lowers the melting point of the aluminium oxide, meaning cheaper energy costs.

Aluminium forms at the negative electrode and oxygen gas forms at the positive carbon electrode. The oxygen reacts with the carbon, forming carbon dioxide.

Key Words   Electrolysis • Current • Electrolyte • Electrode • Reduction • Oxidation • Cryolite

Electrolysis of Sodium Chloride Solution

Sodium chloride (common salt) is a compound of an alkali metal and a halogen. It is found in large quantities in the sea and in underground deposits.

Electrolysis of sodium chloride solution (brine) produces some important reagents for the chemical industry:

- **Chlorine gas** (at the positive **electrode**).
- **Hydrogen gas** (at the negative electrode).
- **Sodium hydroxide solution** (passed out of the cell).

Chlorine is used to kill bacteria in drinking water and swimming pools, and to manufacture hydrochloric acid, disinfectants, bleach and PVC.

Hydrogen is used in the manufacture of ammonia and margarine.

Sodium hydroxide is used in the manufacture of soap, paper and ceramics.

Chlorine bleaches damp litmus paper. This is how its presence can be detected in a laboratory.

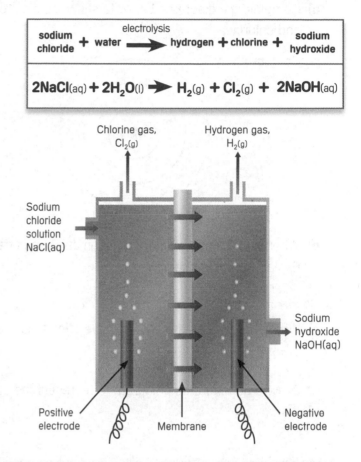

$$\text{sodium chloride} + \text{water} \xrightarrow{\text{electrolysis}} \text{hydrogen} + \text{chlorine} + \text{sodium hydroxide}$$

$$2NaCl_{(aq)} + 2H_2O_{(l)} \rightarrow H_{2(g)} + Cl_{2(g)} + 2NaOH_{(aq)}$$

Chlorine gas, $Cl_{2(g)}$

Hydrogen gas, $H_{2(g)}$

Sodium chloride solution $NaCl_{(aq)}$

Sodium hydroxide $NaOH_{(aq)}$

Positive electrode

Membrane

Negative electrode

HT Electrolysis Equations

Reactions that occur at the electrodes during electrolysis can be represented by **half-equations**.

For example, in the electrolysis of copper...

- copper is deposited at the **negative electrode**

$$Cu^{2+} + 2e^- \rightarrow Cu_{(s)}$$

- chlorine gas is given off at the positive electrode. (Remember that chlorine exists as molecules.)

$$2Cl^- \rightarrow Cl_{2(g)} + 2e^-$$

N.B. When writing equations, remember to include the state symbols.

Quick Test

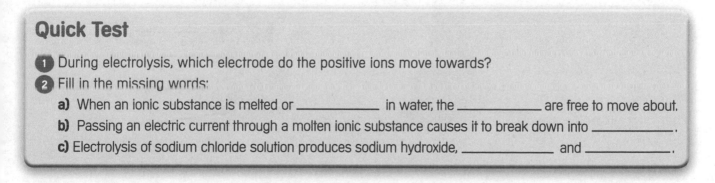

1. During electrolysis, which electrode do the positive ions move towards?
2. Fill in the missing words:
    a) When an ionic substance is melted or _____ in water, the _____ are free to move about.
    b) Passing an electric current through a molten ionic substance causes it to break down into _____ .
    c) Electrolysis of sodium chloride solution produces sodium hydroxide, _____ and _____ .

C2 Exam Practice Questions

1 This question is about the bonding present in sodium chloride (NaCl) and methane (CH_4).

a) Complete the diagrams below to show the electron configurations in an atom of sodium and chlorine. **(2 marks)**

Sodium Atom

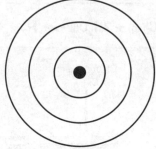

Chlorine Atom

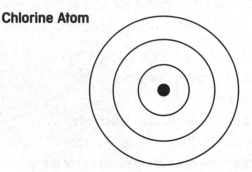

b) When sodium and chlorine react an electron is transferred from the sodium atom to the chlorine atom. Explain why this occurs.

.... **(1 mark)**

c) What charge does the sodium have after it has transferred an electron to the chlorine atom?

.... **(1 mark)**

d) Draw a dot and cross diagram to represent the covalent bonding in a molecule of methane.

(1 mark)

e) Methane has a simple covalent structure. Explain why simple covalent structures have relatively low melting and boiling points.

.... **(1 mark)**

2 An investigation was carried out into the effect of changing concentration on the rate of reaction.

The following apparatus was set up.

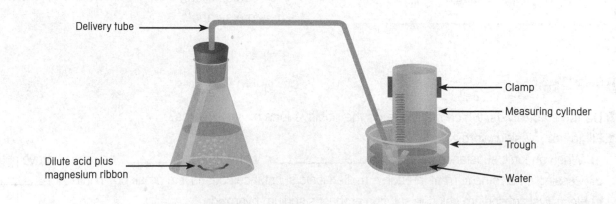

The concentration of acid was chosen. 5cm of magnesium ribbon was added and the time taken for 50cm^3 of gas to be produced was recorded. A different concentration of acid was then used and the experiment repeated.

a) Why is it important that the mass of magnesium ribbon added is the same in each experiment?

.... **(1 mark)**

b) When the concentration of acid is increased how will the time taken to collect 50cm^3 of gas change?

.... **(1 mark)**

c) Explain, in terms of particles, why a higher concentration of acid gives a faster rate of reaction.

....

.... **(1 mark)**

d) Name one other way of increasing the rate of reaction of magnesium with acid.

.... **(1 mark)**

HT **3** Nitric acid (HNO_3) is used to make fertilisers and explosives. Nitric acid reacts with ammonia (NH_3) according to the equation below:

$$HNO_3 + NH_3 \longrightarrow NH_4NO_3$$

a) Calculate the relative formula mass (M_r) of nitric acid.

.... **(1 mark)**

b) Calculate the mass of ammonium nitrate that would be expected to be formed if 6.3g of nitric acid reacts with an excess of ammonia.

....

....

.... **(1 mark)**

c) A student carried out the above experiment but only obtained 3g of ammonium nitrate. Calculate the percentage yield for this reaction. (If you were unable to do part **b)** of this question then use an answer of 5g to help you with part **c)**.

....

....

.... **(1 mark)**

d) Calculate the percentage by mass of nitrogen in ammonium nitrate.

....

....

.... **(1 mark)**

C3 The Periodic Table

Early Attempts to Classify the Elements

Several attempts have been made to group the **elements** in a table, firstly out of curiosity, then as a useful tool to help scientists and finally as an important summary of the structure of atoms.

When **John Newlands** tried to arrange a periodic table in 1864, only 63 elements were known; many were still **undiscovered**. Newlands arranged the known elements in order of their **atomic weights** and found similar properties amongst every eighth element in the series. This makes sense since the noble gases (Group 8) weren't discovered until 1894.

He noticed **periodicity** (repeated patterns) although the missing elements caused problems.

But, strictly following the order of **atomic weight** created problems because it meant some of the elements were placed in the **wrong group**.

Dimitri Mendeleev realised that some elements had yet to be discovered, so when he created his table in 1869 he left **gaps** to allow for their discovery. He used his periodic table to predict the existence of other elements.

The Modern Periodic Table

The discovery of **subatomic particles** (**protons**, **neutrons** and **electrons**) and **electronic structure** early in the 20th century provided further evidence that could be used to create a table. The Periodic Table was then arranged in order of atomic (proton) numbers. So, the modern Periodic Table is an arrangement of the elements in terms of their **electronic structure**.

The elements are arranged in **periods** (rows) according to the **number of electrons** in their outer **energy level** (shell). From left to right across each

period, an energy level is gradually filled with electrons. In the next period, the next energy level is filled, etc.

This arrangement means elements with the same number of electrons in their outer energy level are in the same **group** (column). For example, Group 1 elements have one electron in their outer energy level. Elements that have the **same number of electrons** in their outer energy level have **similar properties**.

The table is called a **Periodic Table** because similar properties occur at **regular intervals**.

1	2										3	4	5	6	7	8 or 0
						1 **H** hydrogen 1										4 **He** helium 2
7 **Li** lithium 3	9 **Be** beryllium 4										11 **B** boron 5	12 **C** carbon 6	14 **N** nitrogen 7	16 **O** oxygen 8	19 **F** fluorine 9	20 **Ne** neon 10
23 **Na** sodium 11	24 **Mg** magnesium 12										27 **Al** aluminium 13	28 **Si** silicon 14	31 **P** phosphorus 15	32 **S** sulfur 16	35.5 **Cl** chlorine 17	40 **Ar** argon 18
39 **K** potassium 19	40 **Ca** calcium 20	45 **Sc** scandium 21	48 **Ti** titanium 22	51 **V** vanadium 23	52 **Cr** chromium 24	55 **Mn** manganese 25	56 **Fe** iron 26	59 **Co** cobalt 27	59 **Ni** nickel 28	63.5 **Cu** copper 29	65 **Zn** zinc 30	70 **Ga** gallium 31	73 **Ge** germanium 32	75 **As** arsenic 33	79 **Se** selenium 34	80 **Br** bromine 35	84 **Kr** krypton 36
85 **Rb** rubidium 37	88 **Sr** strontium 38	89 **Y** yttrium 39	91 **Zr** zirconium 40	93 **Nb** niobium 41	96 **Mo** molybdenum 42	[98] **Tc** technetium 43	101 **Ru** ruthenium 44	103 **Rh** rhodium 45	106 **Pd** palladium 46	108 **Ag** silver 47	112 **Cd** cadmium 48	115 **In** indium 49	119 **Sn** tin 50	122 **Sb** antimony 51	128 **Te** tellurium 52	127 **I** iodine 53	131 **Xe** xenon 54
133 **Cs** caesium 55	137 **Ba** barium 56	139 **La*** lanthanum 57	178 **Hf** hafnium 72	181 **Ta** tantalum 73	184 **W** tungsten 74	186 **Re** rhenium 75	190 **Os** osmium 76	192 **Ir** iridium 77	195 **Pt** platinum 78	197 **Au** gold 79	201 **Hg** mercury 80	204 **Tl** thallium 81	207 **Pb** lead 82	209 **Bi** bismuth 83	[209] **Po** polonium 84	[210] **At** astatine 85	[222] **Rn** radon 86
[223] **Fr** francium 87	[226] **Ra** radium 88	[227] **Ac*** actinium 89	[261] **Rf** rutherfordium 104	[262] **Db** dubnium 105	[266] **Sg** seaborgium 106	[264] **Bh** bohrium 107	[277] **Hs** hassium 108	[268] **Mt** meitnerium 109	[271] **Ds** darmstadtium 110	[272] **Rg** roentgenium 111						

Key Words   Element • Proton • Neutron • Electron • Electronic structure

Group 1 – The Alkali Metals

There are six elements in **Group 1**. They are known as the **alkali metals**.

Alkali metals…

- have **low** melting and boiling points that **decrease** as you go down the group
- have a **low** density (lithium, sodium and potassium are less dense than water)
- become **more reactive** as you go down the group.

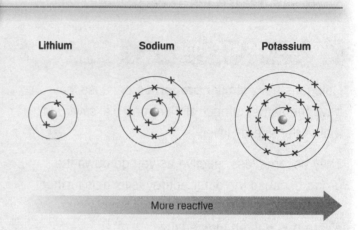

Reactivity increases, and melting and boiling points decrease as you go down the group

HT Trends in Group 1

Alkali metals have **similar properties** to each other because they have the same number of electrons in their outer energy level, i.e. the highest occupied energy level in an atom of each element contains **one electron**.

Alkali metals become **more reactive** as you go down the group because the outer energy level gets further away from the influence of the nucleus, and so an **electron is lost more easily**.

Lithium   Sodium   Potassium

More reactive

Reactions of Alkali Metals

The alkali metals are stored under oil because they react very vigorously with oxygen and water. When alkali metals react with **water**, a **metal hydroxide** is formed and **hydrogen** gas is given off. For example…

potassium	+	water	→	potassium hydroxide	+	hydrogen
$2K_{(s)}$	$+2H_2O_{(l)}$	→	$2KOH_{(aq)}$	$+$	$H_{2(g)}$	

If a metal hydroxide (e.g. potassium hydroxide) is **dissolved** in water, an **alkaline solution** is produced.

Alkali metals react with **non-metals** to form **ionic compounds**. When this happens, the metal atom **loses** one electron to form a metal ion with a **positive charge** (+1). The products are **white solids** that **dissolve** in water to form **colourless** solutions.

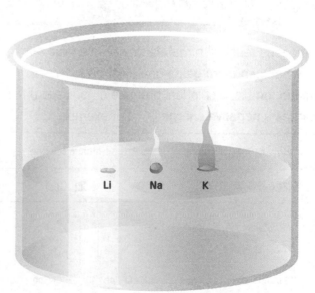

Alkali Metals Reacting with Water

Li   Na   K

C3 The Periodic Table

Group 7 – The Halogens

There are five **elements** in Group 7. They are known as the **halogens**. They are non-metals.

The halogens...
- have melting and boiling points that **increase** as you go down the group (at room temperature, fluorine and chlorine are gases, and bromine is a liquid)
- have **coloured vapours** (chlorine's and bromine's vapours smell particularly strong)
- exist as **molecules** made up of **pairs of atoms**
- become **less reactive** as you go down the group.

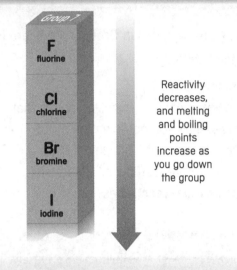

Reactivity decreases, and melting and boiling points increase as you go down the group

HT Trends in Group 7

Halogens have **similar properties** because they have the same number of electrons (i.e. seven) in their outer energy level.

They become less reactive as you go down the group because the outer energy level gets further away from the influence of the nucleus, and so an electron **is gained less easily**.

The more energy levels an atom has...
- the more easily electrons are lost
- the less easily electrons are gained.

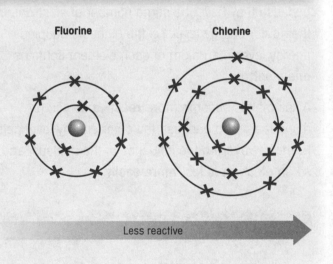

Reactions of Halogens

Halogens react with **metals** to produce **ionic salts**. The **halogen atom gains** one electron to form a **halide ion** (i.e. a chloride, bromide or iodide ion) that carries a **negative charge** (-1). For example:

lithium + chlorine ➡ lithium chloride
$2Li_{(s)}$ + $Cl_2{}_{(g)}$ ➡ $2LiCl_{(s)}$

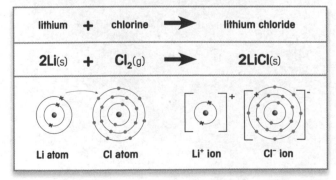

Halogens react with other **non-metallic** elements to form **molecular compounds**. For example:

hydrogen + chlorine ➡ hydrogen chloride
$H_2{}_{(g)}$ + $Cl_2{}_{(g)}$ ➡ $2HCl_{(g)}$

A more reactive halogen will **displace** a less reactive halogen from an aqueous solution of its salt, i.e....
- chlorine will displace both bromine and iodine
- bromine will displace iodine.

Key Words   Halogen • Salt • Compound

The Transition Metals

In the **centre** of the Periodic Table, between Groups 2 and 3, is a block of metallic elements called the **transition metals** (or **transition elements**).

Many transition metals…

- form **coloured compounds**
- have **ions** with **different charges**, e.g. Fe^{2+} and Fe^{3+}
- can be used as **catalysts** to speed up chemical reactions.

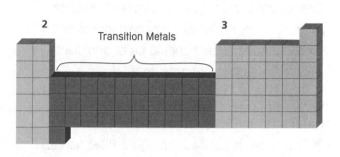

Transition Metals

Properties of Transition Metals

Like all other metals, transition metals…

- are good **conductors** of heat and electricity
- can be **easily bent** or **hammered** into shape.

In comparison to Group 1 metals, transition metals…

- have higher densities and higher melting points (except mercury, which is liquid at room temperature)
- are harder and more mechanically strong (except mercury)
- are much **less reactive** and don't react as vigorously with oxygen or water.

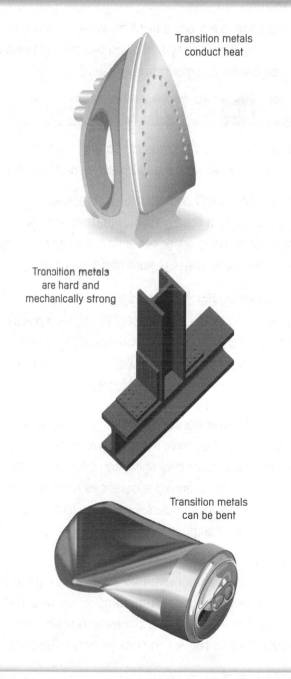

Transition metals conduct heat

Transition metals are hard and mechanically strong

Transition metals can be bent

Quick Test

1. How did Newlands, and then Mendeleev attempt to classify the elements?
2. Upon the discovery of protons, neutrons and electrons, how was the Periodic Table arranged?
3. What gas is produced when lithium reacts with water?
4. Fill in the missing words:
   a) Compared to Group 1 elements, transition metals have _____ melting points, are stronger and_____ and are _____ reactive.
   b) A more reactive halogen can _____ a less reactive halogen from an aqueous solution of its salt.

C3 Water

Drinking Water

Water of the correct quality is **essential** for **life**. Water naturally contains **microorganisms** and **dissolved salts**, which should be of sufficiently low levels to be safe for humans to drink.

Water that is good quality and safe to drink is produced in the following way:

1. The water is passed through a **filter bed** to remove any solid particles.
2. **Chlorine gas** is then added to kill any harmful microorganisms.
3. **Fluoride** is added to drinking water in order to reduce tooth decay (although too much fluoride can cause discolouration of teeth).

To improve the taste and quality of tap water, more dissolved substances can be removed by passing the water through a **filter** containing carbon, silver and **ion exchange resins**.

Any water can be **distilled** to produce **pure water**, i.e. water that contains no dissolved substances. The water is boiled to produce steam, which is condensed by cooling it to produce pure liquid water. This process uses a lot of energy, which makes distillation an expensive process.

Hard and Soft Water

The amount of compounds present in tap water determines whether it's described as **hard** or **soft**.

Soft water doesn't contain many dissolved compounds so it readily forms a **lather** with soap. Permanently hard water remains hard upon boiling. Temporary hard water is softened upon boiling.

(HT) Temporary hard water contains hydrogencarbonate ions (HCO_3^-) that decompose upon heating to produce carbonate (CO_3^{2-}) ions. These carbonate ions react with calcium and/or magnesium ions to form precipitates.

Most **hard water** contains calcium or magnesium compounds that dissolve in natural water as it flows over ground or rocks containing compounds of these elements. These dissolved substances react with soap to form **scum**, which makes it harder to form a lather. Soapless detergents do not form scum.

Advantage of hard water:

- The dissolved compounds in water are **good for your health**, e.g. calcium compounds help the development of strong bones and teeth, and also help to reduce the risk of heart diseases.

Disadvantages of hard water:

- More soap is needed to form a lather, which increases costs.
- Using hard water often leads to deposits (called scale) forming in heating systems and appliances like kettles, which **reduces** their **efficiency**.

Removing Hardness

To make hard water soft, the **dissolved** calcium and magnesium **ions** need to be removed. This can be done in one of two ways:

- Add **sodium carbonate solution** (washing soda) to it. The carbonate ions react with the calcium and magnesium ions to form calcium carbonate and magnesium carbonate (respectively), which precipitate out of solution as they are both insoluble.
- Pass the hard water through an **ion-exchange column**, which contains a resin that supplies hydrogen ions or sodium ions. As the hard water passes through the resin, the calcium and magnesium ions contained in it are replaced by hydrogen or sodium ions from the resin.

Key Words   Ion • Distillation • Efficiency

Calculating and Explaining Energy Change C3

Measuring Energy by Calorimetry

The unit of measurement for **energy** is the **joule** (**J**). It takes 4.2 joules of energy to heat up 1g of water by 1°C. This amount of energy is called 1 **calorie** (**C**), i.e. 1 calorie = 4.2 joules.

Information about the energy provided by food products is given in kilocalories (kcal). When any **chemical change** takes place it is accompanied by an **energy change**, i.e. energy can be taken in or given out. The relative amounts of energy produced by food or fuels can be measured using **calorimetry**.

To measure the temperature change that takes place when a fuel burns, follow this method:

1. Place 100g of water in a calorimeter (a container made of glass or metal) and measure the temperature of the water.
2. Find the mass (in grams) of the fuel to be burned.
3. Burn the fuel under the water in the calorimeter for a few minutes.
4. Record the new temperature and calculate the temperature change of the water.

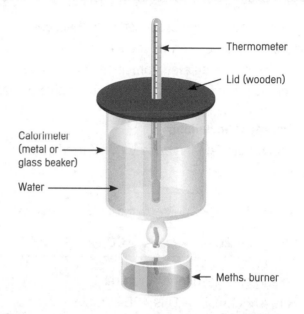

- Thermometer
- Lid (wooden)
- Calorimeter (metal or glass beaker)
- Water
- Meths. burner

5. Weigh the fuel and calculate how much fuel has been used.
6. The energy released (Q) can be calculated by using the following formula:

$$Q = mc\Delta T$$

where m is the mass of water being heated, c is 4.2 (a constant) and ΔT is the temperature change

Making and Breaking Bonds

In a chemical reaction, new substances are produced. In order to do this, the **bonds** in the reactants must be **broken** and new bonds are **made** to form the products.

Breaking a chemical bond requires a lot of energy – this is an **endothermic** process.

When a new chemical bond is **formed**, energy is given out – this is an **exothermic** process.

(HT) If more energy is required to break old bonds than is released when the new bonds are formed, the reaction must be **endothermic**.

If more energy is released when the new bonds are formed than is needed to break the old bonds, the reaction must be **exothermic**.

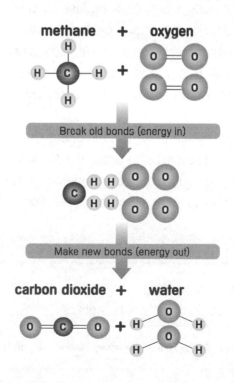

methane + oxygen

Break old bonds (energy in)

Make new bonds (energy out)

carbon dioxide + water

C3 Calculating and Explaining Energy Changes

Measuring Energy of Chemical Reactions

The amount of energy produced in a **chemical reaction** in solution can be measured by mixing the reactants in an **insulated** container. This enables the temperature change to be measured before heat is lost to the surroundings. This method would be suitable for **neutralisation** reactions and reactions involving solids, e.g. zinc and acid.

HT Energy Calculations

Example: Calculate the energy transferred in the following reaction:

methane **+** oxygen $\longrightarrow$ carbon dioxide **+** water
$CH_{4(g)}$ **+** $2O_{2(g)}$ $\longrightarrow$ $CO_{2(g)}$ **+** $2H_2O_{(l)}$

Bond energies needed for this are:
C–H is 412kJ/mol, O=O is 496kJ/mol,
C=O is 805kJ/mol, H–O is 463kJ/mol

Energy used to break bonds is:
4 C–H = 4 x 412 = 1648kJ, 2 O=O = 2 x 496 = 992kJ
Total = 1648kJ + 992kJ = **2640kJ**

Energy given out by making bonds:
2 C=O = 2 x 805 = 1610kJ
4 H–O = 4 x 463 = 1852kJ
Total = 1610kJ + 1852kJ = **3462kJ**

Energy change (ΔH) = Energy used to break bonds – Energy given out by making bonds

= 2640kJ – 3462kJ = **–822kJ**

Energy Level Diagrams

The energy changes in a chemical reaction can be illustrated using an **energy level diagram**:

1. In an **exothermic** reaction, energy is given out. This means energy is being lost, so the products have less energy than the reactants.
2. In an **endothermic** reaction, energy is being taken in. This means that energy is being gained, so the products have more energy than the reactants.
3. The **activation energy** is the energy needed to start a reaction, i.e. to break the old bonds.
4. **Catalysts** reduce the activation energy needed for a reaction – this makes the reaction go faster.

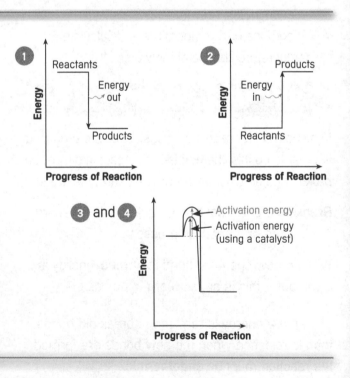

Hydrogen as a Fuel

Hydrogen can be used…
- as a fuel in combustion engines
- in fuel cells, which produce electricity that can be used to power vehicles.

hydrogen **+** oxygen $\longrightarrow$ water **+** heat

Neutralisation • Activation energy

Flame Tests

Flame tests can be used to identify metal **ions**.

Lithium, sodium, potassium, calcium and barium compounds can be recognised by the distinctive colours they produce in a **flame test**.

To do a flame test, follow this method:

1. Heat and then dip a piece of nichrome (a nickel-chromium alloy) wire in concentrated hydrochloric acid to clean it.
2. Dip the wire in the compound.
3. Put it into a Bunsen flame. The following distinctive colours indicate the presence of certain ions:
   - Green for **barium**
   - Brick red for **calcium**
   - Crimson red for **lithium**
   - Lilac for **potassium**
   - Yellow for **sodium**.

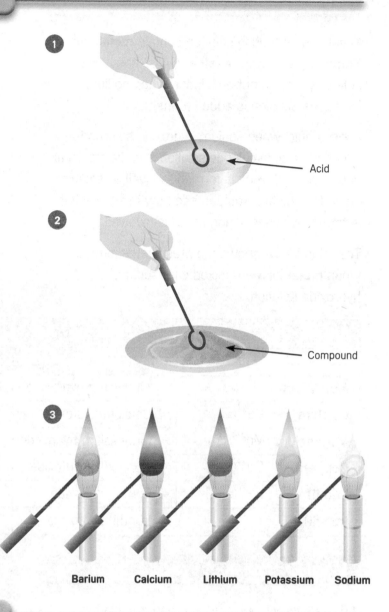

Acid

Compound

Barium   Calcium   Lithium   Potassium   Sodium

Reacting Carbonates with Dilute Acid

Carbonates react with **dilute acids** to form **carbon dioxide** gas (and a salt and water). Carbon dioxide turns limewater milky. For example:

calcium carbonate	+	hydrochloric acid	→	calcium chloride	+	carbon dioxide	+	water

$$CaCO_3{}_{(s)} + 2HCl_{(aq)} \longrightarrow CaCl_2{}_{(aq)} + CO_2{}_{(g)} + H_2O_{(l)}$$

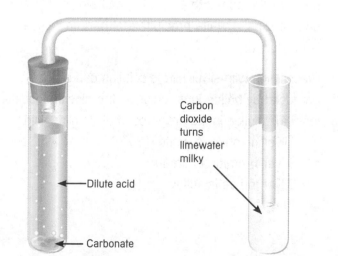

Carbon dioxide turns limewater milky

Dilute acid

Carbonate

Precipitation of Metal Ions

Metal compounds in solution contain **metal ions**. Some of these form **precipitates**, i.e. **insoluble** solids that come out of solution when sodium hydroxide solution is added to them.

For example, when sodium hydroxide solution is added to calcium chloride solution, a white precipitate of calcium hydroxide is formed (as well as sodium chloride solution). You can see how this precipitate is formed by considering the ions involved.

The table below shows the precipitates formed when **metal ions** are mixed with sodium hydroxide solution.

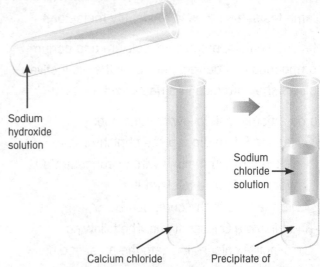

Sodium hydroxide solution

Sodium chloride solution

Calcium chloride solution

Precipitate of calcium hydroxide

Metal Ion		Precipitate Formed	
		Precipitate	Precipitate Colour
Aluminium	Al^{3+}(aq)	Aluminium hydroxide	White (dissolves with excess sodium hydroxide)
Calcium	Ca^{2+}(aq)	Calcium hydroxide	White
Magnesium	Mg^{2+}(aq)	Magnesium hydroxide	White
Copper(II)	Cu^{2+}(aq)	Copper(II) hydroxide	Blue
Iron(II)	Fe^{2+}(aq)	Iron(II) hydroxide	Green
Iron(III)	Fe^{3+}(aq)	Iron(III) hydroxide	Brown

More Examples of Precipitations

If dilute hydrochloric acid and barium chloride solution are added to a solution containing **sulfate ions**, a white precipitate of barium sulfate is produced.

Precipitates with silver nitrate solution can be produced by **halide ions** (chloride, bromide and iodide ions) in solution in the presence of dilute nitric acid:
- Silver chloride is white.
- Silver bromide is cream.
- Silver iodide is yellow.

Quick Test

1. What is produced when hard water reacts with soap?
2. What are the normal units of energy?
3. What happens to energy when bonds are formed?
4. In a flame test, which compound results in a crimson flame?
5. What colour precipitate do calcium ions form with a sodium hydroxide solution?
6. What kind of ions form a blue precipitate with sodium hydroxide solution?
7. What kind of precipitate do chloride ions form with silver nitrate solution in the presence of nitric acid?

Titration

Titration is an accurate technique that you can use to find out **how much** of an **acid** is needed to **neutralise** an alkali.

When **neutralisation** takes place, the hydrogen ions (H^+) from the acid join with the hydroxide ions (OH^-) from the alkali to form water (neutral pH).

hydrogen ion	+	hydroxide ion	→	water molecule
$H^+_{(aq)}$	+	$OH^-_{(aq)}$	→	$H_2O_{(l)}$

Use this titration method:

1. Wash and rinse a pipette with the alkali that you will use.
2. Use the pipette to measure out a known and accurate volume of the alkali.
3. Place the alkali in a clean, dry conical flask. Add a suitable indicator, e.g. phenolphthalein.
4. Place the acid in a burette that has been carefully washed and rinsed with the acid. Take a reading of the volume of acid in the burette (initial reading).
5. Carefully add the acid to the alkali until the indicator changes colour to show neutrality. This is called the **end point**. Take a reading of the volume of acid in the burette (final reading).
6. Calculate the volume of acid added (i.e. subtract the final reading from the initial reading).

This method can be repeated to check results and can then be performed without an indicator in order to obtain the salt.

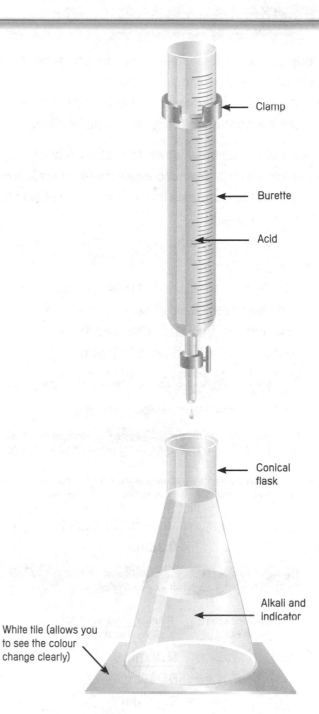

Clamp

Burette

Acid

Conical flask

Alkali and indicator

White tile (allows you to see the colour change clearly)

Indicators

Different strength acids and alkalis can react together to form a neutral solution. You must use a suitable **indicator** in titrations. For example, if you have a strong acid and strong alkali you should use any suitable acid–base indicator (e.g. litmus).

Universal Indicator Solution

⊞ Titration

Titration can be used to find the concentration of an acid or alkali providing you know either…
- the relative volumes of acid and alkali used or
- the concentration of the other acid or alkali.

It will help if you break down the calculation:

① Write down a balanced equation for the reaction in order to determine the ratio of moles of acid to alkali involved.

② Calculate the number of moles in the solution of known volume and concentration. (You will know the number of moles in the other solution from your previous calculation.)

③ Calculate the concentration of the other solution using this formula:

$$\text{Concentration of solution (mol dm}^{-3}\text{ or M)} = \frac{\text{Number of moles of solute (mol)}}{\text{Volume of solution (dm}^{-3}\text{)}}$$

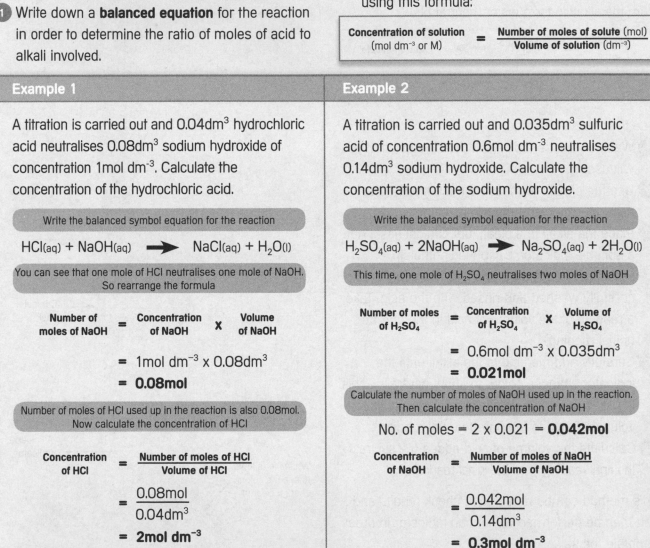

Example 1

A titration is carried out and 0.04dm³ hydrochloric acid neutralises 0.08dm³ sodium hydroxide of concentration 1mol dm⁻³. Calculate the concentration of the hydrochloric acid.

Write the balanced symbol equation for the reaction

$$HCl_{(aq)} + NaOH_{(aq)} \longrightarrow NaCl_{(aq)} + H_2O_{(l)}$$

You can see that one mole of HCl neutralises one mole of NaOH. So rearrange the formula

$$\text{Number of moles of NaOH} = \text{Concentration of NaOH} \times \text{Volume of NaOH}$$
$$= 1\text{mol dm}^{-3} \times 0.08\text{dm}^3$$
$$= \textbf{0.08mol}$$

Number of moles of HCl used up in the reaction is also 0.08mol. Now calculate the concentration of HCl

$$\text{Concentration of HCl} = \frac{\text{Number of moles of HCl}}{\text{Volume of HCl}}$$
$$= \frac{0.08\text{mol}}{0.04\text{dm}^3}$$
$$= \textbf{2mol dm}^{-3}$$

Example 2

A titration is carried out and 0.035dm³ sulfuric acid of concentration 0.6mol dm⁻³ neutralises 0.14dm³ sodium hydroxide. Calculate the concentration of the sodium hydroxide.

Write the balanced symbol equation for the reaction

$$H_2SO_{4(aq)} + 2NaOH_{(aq)} \longrightarrow Na_2SO_{4(aq)} + 2H_2O_{(l)}$$

This time, one mole of H₂SO₄ neutralises two moles of NaOH

$$\text{Number of moles of H}_2\text{SO}_4 = \text{Concentration of H}_2\text{SO}_4 \times \text{Volume of H}_2\text{SO}_4$$
$$= 0.6\text{mol dm}^{-3} \times 0.035\text{dm}^3$$
$$= \textbf{0.021mol}$$

Calculate the number of moles of NaOH used up in the reaction. Then calculate the concentration of NaOH

$$\text{No. of moles} = 2 \times 0.021 = \textbf{0.042mol}$$

$$\text{Concentration of NaOH} = \frac{\text{Number of moles of NaOH}}{\text{Volume of NaOH}}$$
$$= \frac{0.042\text{mol}}{0.14\text{dm}^3}$$
$$= \textbf{0.3mol dm}^{-3}$$

Quick Test

① What is the name of the technique used to accurately determine how much acid is needed to neutralise an alkali?

② Name a suitable indicator for accurately neutralising a strong alkali with a strong acid.

③ What is the concentration of a solution of sodium hydroxide that contains three moles in 4dm³?

The Haber Process

Reversible reactions may not go to completion. But they can still be used efficiently in continuous processes such as the **Haber process.**

The Haber process is used to manufacture **ammonia**. The raw materials for this process are...
- **nitrogen** – from the fractional distillation of liquid air
- **hydrogen** – from natural gas and steam.

The purified nitrogen and hydrogen are passed over an **iron catalyst** at a...
- **high temperature** (about 450°C)
- **high pressure** (about 200 atmospheres).

Some of the hydrogen and nitrogen **reacts** to **form ammonia**. The **ammonia** produced can **break down** again into **nitrogen** and **hydrogen**.

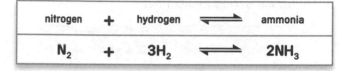

nitrogen	+	hydrogen	⇌	ammonia
N_2	+	$3H_2$	⇌	$2NH_3$

(HT) These reaction conditions are chosen to produce a **reasonable yield** of ammonia quickly.

Even so, only **some** of the hydrogen and nitrogen react together to form ammonia.

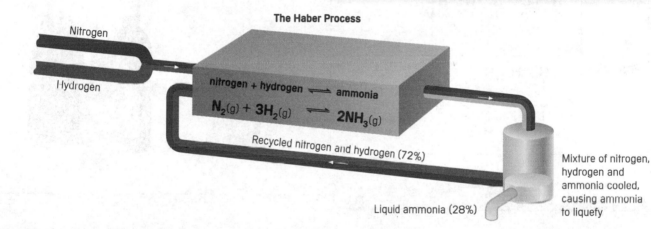

The Haber Process

Nitrogen

Hydrogen

nitrogen + hydrogen ⇌ ammonia

$N_2(g) + 3H_2(g)$ ⇌ $2NH_3(g)$

Recycled nitrogen and hydrogen (72%)

Liquid ammonia (28%)

Mixture of nitrogen, hydrogen and ammonia cooled, causing ammonia to liquefy

(HT) Closed Systems

In a **closed system**, no reactants are added and no products are removed. When a reversible reaction occurs in a closed system, an **equilibrium** is achieved when the reactions occur at exactly the **same rate** in **each direction**. The relative amounts of all the reacting substances at equilibrium depend on the conditions of the reaction.

(HT) Changing Reaction Conditions

In an **exothermic** reaction...
- if the temperature is **raised**, the **yield decreases**
- if the temperature is **lowered**, the yield **increases**.

In an **endothermic** reaction...
- if the temperature is **raised**, the yield **increases**
- if the temperature is **lowered**, the yield **decreases**.

In **gaseous reactions**, an increase in pressure favours the reaction that produces the least number of molecules.

These factors, together with reaction rates, determine the optimum conditions in industrial processes, e.g. the **Haber process**.

C3 Alcohols, Carboxylic Acids and Esters

Alcohols

Alcohols are carbon-based molecules that contain the **functional group** –OH. Methanol, ethanol and propanol are the first three members of the **homologous series** of alcohols.

Alcohol	Structural Formula	Formula
Methanol	H | H — C — O — H | H	CH_3OH
Ethanol	H H | | H — C — C — O — H | | H H	CH_3CH_2OH
Propanol	H H H | | | H — C — C — C — O — H | | | H H H	$CH_3CH_2CH_2OH$

Alcohols…
- dissolve in water to form neutral solutions
- react with sodium to produce hydrogen
- burn in air
- are used as fuels and **solvents**.

Alcoholic drinks contain ethanol. Ethanol can be **oxidised** to ethanoic acid either by chemical oxidising agents or by **microbial** action.

Ethanoic acid is the main acid in vinegar.

Carboxylic Acids

Carboxylic acids are carbon-based molecules that contain the functional group –COOH.

Carboxylic Acid	Structural Formula	Formula
Methanoic acid	O || C / \\ H OH	COOH
Ethanoic acid	H O | || H — C — C | \\ H O — H	CH_3COOH
Propanoic acid	H H O | | || H — C — C — C | | \\ H H O — H	CH_3CH_2COOH

Carboxylic acids…
- dissolve in water to form acidic solutions
- react with carbonates (e.g. sodium carbonate) to produce carbon dioxide
- react with alcohols (in the presence of an acid catalyst) to form esters.

(HT) Carboxylic acids don't ionise (dissociate) fully in water, so they are called weak acids.

Aqueous solutions of weak acids have a higher pH than aqueous solutions of strong acids with the same concentration.

Esters

Alcohols and carboxylic acids react together to form **esters**. When ethanol and ethanoic acid react together the ester formed is ethyl ethanoate.

Esters contain the functional group −COO as shown below:

Ethyl ethanoate (CH$_3$COOC$_2$H$_5$)

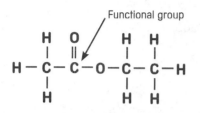

Esters are **volatile** compounds (meaning they have a low boiling point).

They have distinctive smells, so are used in perfumes and as flavourings in food.

Quick Test

1. What are the raw materials in the Haber process?
2. What catalyst is used in the Haber process?
3. In the manufacture of ammonia, what temperature and pressure is used?
4. What functional group do all alcohols contain?
5. Which of the following statements is incorrect?
   a) Alcohols dissolve in water to form a neutral solution.
   b) Alcohols react with sodium to produce carbon dioxide.
   c) Alcohols burn in air.
6. What is the name of the carboxylic acid shown below?
7. Give one use of esters.

C3 Exam Practice Questions

1 This question is about elements in Group 1 and Group 7.

a) What name is given to the group of metals in the Periodic Table that contains lithium?

.... **(1 mark)**

b) At the end of each of the statements below state whether they are true or false.

i) Alkali metals have low melting points that increase as you go down

the group. .... **(1 mark)**

ii) Alkali metals are less dense than water. .... **(1 mark)**

iii) Alkali metals become more reactive as you go down the group. .... **(1 mark)**

iv) Lithium fluoride will be a white solid .... **(1 mark)**

v) Lithium fluoride will be insoluble in water .... **(1 mark)**

c) Fluorine is the most reactive element in Group 7. It reacts with sodium chloride as shown below:

fluorine + sodium chloride ⟶ sodium fluoride + chlorine

What type of reaction is this?

.... **(1 mark)**

2 An energy level diagram for a reaction is shown below.

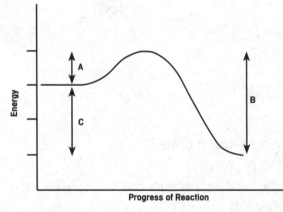

a) Is this reaction exothermic or endothermic? .... **(1 mark)**

b) Explain, in terms of making and breaking bonds, your answer to part **a**.

....

.... **(1 mark)**

c) Which arrow, A, B or C, represents the activation energy for the reaction?

.... **(1 mark)**

d) What is the effect of adding a catalyst on the activation energy for this reaction?

.... **(1 mark)**

3 Ammonia gas (NH_3) is formed by the Haber process. The equation for the reaction is shown below:

$$N_2(g) + 3H_2(g) \rightleftharpoons 2NH_3(g)$$

a) What is meant by the $\rightleftharpoons$ symbol?

.... **(1 mark)**

b) What temperature is used in the Haber process?

.... **(1 mark)**

c) The production of ammonia is exothermic. What is the effect on the yield of ammonia if the temperature is increased?

.... **(1 mark)**

d) What pressure is used in the Haber process?

.... **(1 mark)**

e) State one use of ammonia.

.... **(1 mark)**

4 A student was carrying out a titration to determine the concentration of a solution of sodium hydroxide. She placed 25cm³ of sodium hydroxide into a conical flask and added five drops of indicator. She then carried out the titration using 1mol dm⁻³ hydrochloric acid. The volume of hydrochloric acid used was 22.50cm³.

The equation for the reaction is shown below:

$$NaOH + HCl \longrightarrow NaCl + H_2O$$

a) What piece of apparatus was used to add the acid to the sodium hydroxide?

.... **(1 mark)**

b) Why was an indicator added?

.... **(1 mark)**

c) The number of moles of hydrochloric acid reacting in this titration was 0.0225. How many moles of sodium hydroxide reacted? Explain your answer.

....

.... **(2 marks)**

d) Calculate the concentration of the sodium hydroxide solution.

Use the equation: Concentration of solution = $\dfrac{\text{Number of moles of solute}}{\text{Volume of solution}}$

.... **(1 mark)**

Answers

Unit 1

Quick Test Answers
Page 14
1. Protons and neutrons
2. The number of electrons in the highest energy level of an atom is the same as the group the element is in, e.g. all atoms in Group 2 have two electrons in the highest energy level. (Helium is the only exception).
3. 2,8,2

Page 17
1. Calcium oxide and carbon dioxide
2. Carbon dioxide turns limewater cloudy
3. (Powdered) clay
4. **a)** $4Al + 3O_2 \rightarrow 2Al_2O_3$
   b) $2Cr + 6HCl \rightarrow 2CrCl_3 + 3H_2$

Page 20
1. concentrated; purified
2. Carbon
3. Phytomining and bioleaching
4. A mixture of a metal with at least one other substance/element (e.g. carbon or another metal)
5. Because the layers of atoms are easily able to slide over each other
6. To strengthen them
7. Transition metals

Page 25
1. Carbon and hydrogen
2. Fractional distillation
3. **Any two from:** Carbon dioxide; Water (vapour); Carbon monoxide; Sulfur dioxide; Oxides of nitrogen
4. Acid rain and global warming
5. Cracking
6. Alkenes
7. Add bromine water; it turns from orange to colourless (is decolourised).
8. Monomers

Page 27
1. They provide energy and nutrients
2. Vegetable oils have a higher boiling point than water

3. **Any two from:** Salad dressings; Ice creams; Cosmetics; Paints
4. Double carbon–carbon bonds (C = C)

Page 29
1. **a)** mantle, crust
   b) move
   c) tectonic plates

Page 31
1. **a)** Nitrogen and oxygen
   b) **Any one from:** Carbon dioxide; Water vapour; Noble gases
2. By plants (photosynthesis)
3. In sedimentary rocks and by the oceans

Exam Practice Answers
1. **a)** Positive
   b) 11
   c)
   d) **i)** 3
      ii) $4Na + O_2 \rightarrow 2Na_2O$
2. **a)** Quarried / mined
   b) **Any one from:** Magnesium; Copper; Zinc; Calcium; Sodium
   c) Limewater turns milky (or cloudy)
3. **a)** It is a good conductor of electricity
   b) By heating the ore in a furnace (smelting)
   c) Electrolysis
   d) **Any one from:** Phytomining; Bioleaching
4. **a)** Two or more elements or compounds that are not chemically combined
   b) Carbon and hydrogen
   c) **Any one from:** Carbon dioxide; Carbon monoxide; Water vapour; Sulfur dioxide
   d) Cracking
5. **a)** X   **b)** Y   **c)** Y   **d)** Y

Unit 2

Quick Test Answers
Page 35
1. Two or more atoms chemically combined together
2. Covalent
3. **a)** melting points / boiling points; intermolecular
   b) electricity; move

Page 39
1. The layers of atoms are able to slide over each other.
2. Thermo-setting polymers consist of cross-links between the polymer chains. Thermo-softening polymers do not.
3. **Any two from:** In computers; Catalysts; Coatings; Highly selective sensors; Stronger and lighter construction materials; New cosmetics, e.g. suntan creams and deodorants

Page 41
1. Mass number
2. Isotopes

Page 43
1. **a)** 22%
   b) 35%
   c) 21%

2. **a)** 2
   b) 0.19
3. **a)** 56g
   b) 414g

Page 46
1. **Any two from:** Instrumental methods are accurate; Sensitive and rapid; Useful when dealing with small quantities
2. **Any two from:** The reaction may be reversible; Some of the product may be lost when it's separated from the reaction mixture; Some of the reactants may react in ways different to the expected reaction.
3. The relative molecular mass of a compound
4. 11g

Page 50
1. Particles are given more energy so there will be more collisions and more energetic collisions (i.e. more collisions where the particles collide with sufficient energy to react).
2. increase; reduce / decrease
3. The surroundings
4. It will be endothermic

Unit 2 (Cont.)

Page 53
1. Precipitation reactions
2. **a)** Alkalis
   b) Fertilisers
   c) Water

Page 55
1. Negative (cathode)
2. **a)** dissolved; ions
   b) elements
   c) hydrogen; chlorine

Exam Practice Answers
1. **a)**
   Sodium Atom   **Chlorine Atom**

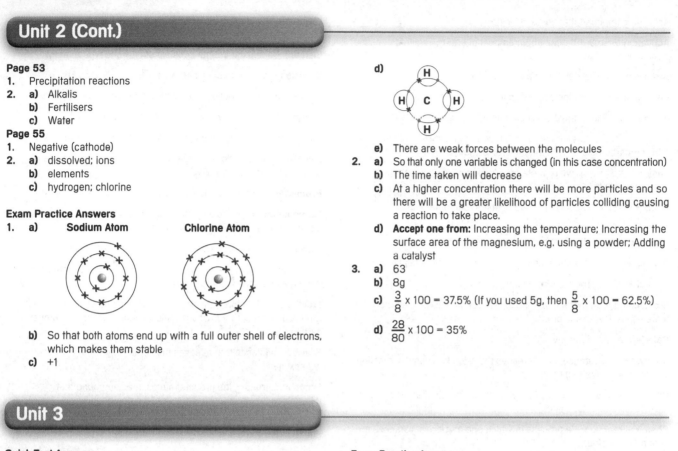

   b) So that both atoms end up with a full outer shell of electrons, which makes them stable
   c) +1

d)

e) There are weak forces between the molecules
2. **a)** So that only one variable is changed (in this case concentration)
   b) The time taken will decrease
   c) At a higher concentration there will be more particles and so there will be a greater likelihood of particles colliding causing a reaction to take place.
   d) **Accept one from:** Increasing the temperature; Increasing the surface area of the magnesium, e.g. using a powder; Adding a catalyst
3. **a)** 63
   b) 8g
   c) $\frac{3}{8}$ x 100 = 37.5% (If you used 5g, then $\frac{5}{8}$ x 100 = 62.5%)
   d) $\frac{28}{80}$ x 100 = 35%

Unit 3

Quick Test Answers
Page 61
1. In terms of their atomic weight
2. In order of atomic number
3. Hydrogen
4. **a)** higher; harder; less
   b) displace

Page 66
1. Scum
2. Joules
3. It's released
4. Lithium
5. White
6. Copper(II)
7. White

Page 68
1. A titration
2. Litmus
3. 0.75mol dm⁻³

Page 71
1. Nitrogen and hydrogen
2. Iron
3. 450°C and 200 atmospheres
4. −OH
5. b)
6. Ethanoic acid
7. **Accept one from:** Perfumes; Food flavourings

Exam Practice Answers
1. **a)** Alkali metals
   b) **i)** False
       ii) True
       iii) True
       iv) True
       v) False
   c) Displacement / Redox
2. **a)** Exothermic
   b) More energy is released in making new bonds than is used in breaking the bonds in the reactants
   c) A
   d) The activation energy is lowered
3. **a)** Reversible reaction (or equilibrium)
   b) 450°C
   c) The yield (amount) of ammonia decreases
   d) 200 atmospheres
   e) **Accept one from:** Manufacture of nitric acid; Manufacture of fertilisers
4. **a)** Burette
   b) To know when neutralisation has occurred / The end-point has been reached
   c) 0.0225; The NaOH and HCl react in a 1 to 1 ratio
   d) 0.9mol dm⁻³

Glossary

Acid – a compound that has a pH lower than 7

Activation energy – the minimum amount of energy required to cause a reaction

Alkali – a compound that has a pH higher than 7

Alkali metal – one of the six metals in Group 1 of the Periodic Table

Alkane – a saturated hydrocarbon with the general formula C_nH_{2n+2}

Alkene – an unsaturated hydrocarbon (with at least one double carbon–carbon bond) with the general formula C_nH_{2n}

Alloy – a mixture of two or more metals, or a mixture of one metal and a non-metal

Atom – the smallest part of an element that can enter into a chemical reaction

Atomic number – the number of protons in an atom

Biodegradable – a substance that does decompose naturally

Biofuel – a fuel produced from plant material

Bioleaching – an extraction method that uses bacteria to extract metals from low-grade ores

Calorie – a unit of energy

Catalyst – a substance that increases the rate of a chemical reaction without being changed itself

Chemical formula – a way of showing the elements that are present in molecules of a substance

Chemical reaction – a process in which one or more substances are changed into others

Chromatography – a technique used to separate different compounds in a mixture according to how well they dissolve a particular solvent

Compound – a substance consisting of two or more elements chemically combined together

Conductor – a substance that readily transfers heat or energy

Covalent bond – a bond between two atoms, in which both atoms share one or more electrons

Cracking – the process used to break down long-chain hydrocarbons into more useful short-chain hydrocarbons, using high temperatures and a catalyst

Crude oil – a liquid mixture found in rocks that contains hydrocarbons

Cryolite – an aluminium-containing ionic compound; used in the electrolytic recovery of aluminium

Current – the flow of electric charge through a conductor

Decomposition – breaking down

Distillation – a process of separating a liquid mixture by boiling it and condensing its vapours

Efficiency – the energy output expressed as a percentage of energy input

Electrode – a piece of metal or carbon that allows electric current to enter and leave during electrolysis

Electrolysis – the process by which an electric current causes a solution to undergo chemical decomposition

Electrolyte – the molten or aqueous solution of an ionic compound used in electrolysis

Electron – a negatively charged subatomic particle that orbits the nucleus

Electronic structure – the arrangement of electrons around the nucleus of an atom

Element – a substance that consists of only one type of atom

Emulsion – a mixture of oil and water

Endothermic – a reaction that takes in heat from its surroundings

Energy – the ability to do work; measured in joules (J)

Ester – organic compound containing the functional group –COO

Evidence – observations, measurements and data collected and subjected to some form of validation

Exothermic – a reaction that gives out heat to its surroundings

Fermentation – the process by which yeast converts sugars to alcohol and carbon dioxide through anaerobic respiration

Fossil – the remains of animals / plants preserved in rock

Fossil fuel – fuel formed in the ground, over millions of years, from the remains of dead plants and animals

Fractional distillation – the process used to separate crude oil into groups of hydrocarbons whose molecules have a similar number of carbon atoms

Fuel – a substance that releases heat or energy when combined with oxygen

Functional group – the group of atoms in a compound that determines the chemical behaviour of the compound

Global warming – the increase in the average temperature on Earth due to a rise in the levels of greenhouse gases in the atmosphere

Halogen – one of the five non-metals in Group 7 of the Periodic Table

Homologous series – a series of organic compounds with the same general formula that have similar chemical properties

Hydrocarbon – a compound containing only hydrogen and carbon

Insoluble – a substance that will not dissolve in a solvent

Ion – a charged particle formed when an atom gains or loses electrons

Ionic bond – the bond formed between two (or more) atoms when one loses, and another gains, electrons to become charged ions

Ionic compound – a compound formed when two (or more) elements bond ionically

Isotopes – atoms of the same element that contain different numbers of neutrons

Joule (J) – a unit of energy

Mass number – the total number of protons and neutrons present in an atom

Microbial – relating to microbes

Mineral – a naturally occurring chemical element or compound found in rocks

Mixture – two or more elements or compounds that are not chemically combined

Mole (mol) – the molar mass of a substance, i.e. the mass in grams of 6×10^{23} particles

Monomer – a small hydrocarbon molecule containing a double bond

Nanoscience – dealing with materials that have a very small grain size, in the order of 1–100nm

Neutralisation – a reaction between an acid and a base that forms a neutral solution (i.e. pH 7)

Neutralise – to form a neutral solution

Neutron – a subatomic particle found in the nucleus of an atom that has no charge

Non-biodegradable – a substance that doesn't decompose naturally by the action of microorganisms

Nucleus – the small central core of an atom, consisting of protons and neutrons

Ore – a naturally occurring mineral, from which it is economically viable to extract

Oxidation – a reaction involving the gain of oxygen, the loss of hydrogen, or the loss of electrons

Oxidised – a substance that gained oxygen and / or lost electrons

pH – a measure of acidity or alkalinity

Phytomining – an extraction method that uses plants to extract copper

Pollution – the contamination of an environment by chemicals, waste or heat

Polymer – a giant, long-chained hydrocarbon

Polymerisation – the process of monomers joining together to form a polymer

Precipitate – an insoluble solid formed in a precipitation reaction

Precipitation – a type of reaction in which a solid is made when two liquids are mixed

Product – a substance made at the end of a chemical reaction

Proton – a positively charged subatomic particle found in the nucleus

Reactant – a substance present before a chemical reaction takes place

Reduction – a reaction involving the loss of oxygen, the gain of hydrogen, or the gain of electrons

Relative atomic mass (A_r) – the average mass of an atom of an element compared with a twelfth of the mass of a carbon atom.

Relative formula mass (M_r) – the sum of the atomic masses of all atoms in a molecule

Reversible reaction – a reaction in which products can react to re-form the original reactants

Salt – the product of a chemical reaction between a base and an acid

Saturated (hydrocarbon) – a hydrocarbon molecule with no double bonds

Sedimentary rock – rock formed by the accumulation of sediment

Smart alloy – an alloy that can change shape and then return to its original shape

Smelting – a method of extracting a metal from its ore by heating the ore in a furnace

Soluble – a substance that can dissolve in a solvent

Solvent – the substance that dissolves the solute

Tectonic plates – huge sections of the Earth's crust that move in relation to one another

Theory – the best way to explain why something is happening. It can be changed when new evidence is found

Thermal decomposition – the breakdown of a chemical substance due to the action of heat

Thermo-setting (polymer) – polymer chains that are joined together by cross-links

Thermo-softening (polymer) – polymer chains that are tangled together

Titration – a method used to find the concentration of an acid or alkali

Unsaturated – a term used to describe alkenes that identifies the presence of a C=C bond

Yield – the amount of a product obtained from a reaction

HT Emulsifier – a substance that helps to stabilise an emulsion

Equilibrium – the state in which a chemical reaction proceeds at the same rate as its reverse reaction (the reactants are balanced)

Hydrogenation – the process in which hydrogen is used to harden vegetable oils

Hydrophilic – water-loving molecule

Hydrophobic – water-hating molecule

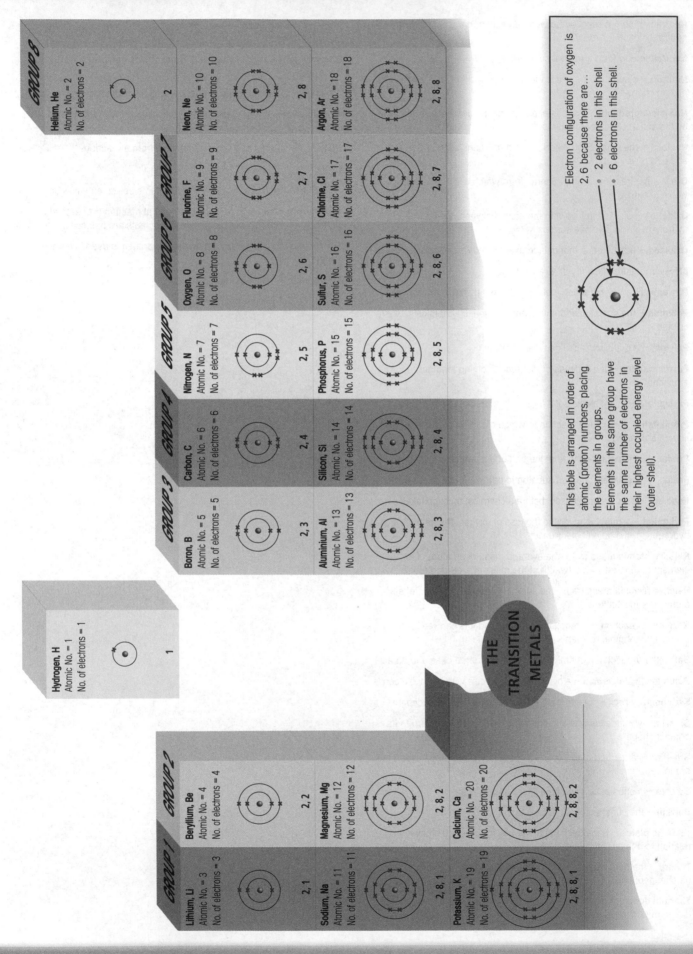

Hydrogen, H
Atomic No. = 1
No. of electrons = 1

1

GROUP 1

Lithium, Li
Atomic No. = 3
No. of electrons = 3

2, 1

Sodium, Na
Atomic No. = 11
No. of electrons = 11

2, 8, 1

Potassium, K
Atomic No. = 19
No. of electrons = 19

2, 8, 8, 1

GROUP 2

Beryllium, Be
Atomic No. = 4
No. of electrons = 4

2, 2

Magnesium, Mg
Atomic No. = 12
No. of electrons = 12

2, 8, 2

Calcium, Ca
Atomic No. = 20
No. of electrons = 20

2, 8, 8, 2

THE TRANSITION METALS

GROUP 3

Boron, B
Atomic No. = 5
No. of electrons = 5

2, 3

Aluminium, Al
Atomic No. = 13
No. of electrons = 13

2, 8, 3

GROUP 4

Carbon, C
Atomic No. = 6
No. of electrons = 6

2, 4

Silicon, Si
Atomic No. = 14
No. of electrons = 14

2, 8, 4

GROUP 5

Nitrogen, N
Atomic No. = 7
No. of electrons = 7

2, 5

Phosphorus, P
Atomic No. = 15
No. of electrons = 15

2, 8, 5

GROUP 6

Oxygen, O
Atomic No. = 8
No. of electrons = 8

2, 6

Sulfur, S
Atomic No. = 16
No. of electrons = 16

2, 8, 6

GROUP 7

Fluorine, F
Atomic No. = 9
No. of electrons = 9

2, 7

Chlorine, Cl
Atomic No. = 17
No. of electrons = 17

2, 8, 7

GROUP 8

Helium, He
Atomic No. = 2
No. of electrons = 2

2

Neon, Ne
Atomic No. = 10
No. of electrons = 10

2, 8

Argon, Ar
Atomic No. = 18
No. of electrons = 18

2, 8, 8

Electron configuration of oxygen is 2, 6 because there are...
- 2 electrons in this shell
- 6 electrons in this shell.

This table is arranged in order of atomic (proton) numbers, placing the elements in groups.
Elements in the same group have the same number of electrons in their highest occupied energy level (outer shell).

Key

relative atomic mass
atomic symbol
name
atomic (proton) number

1	2												3	4	5	6	7	0
						1 **H** hydrogen 1												4 **He** helium 2
7 **Li** lithium 3	9 **Be** beryllium 4												11 **B** boron 5	12 **C** carbon 6	14 **N** nitrogen 7	16 **O** oxygen 8	19 **F** fluorine 9	20 **Ne** neon 10
23 **Na** sodium 11	24 **Mg** magnesium 12												27 **Al** aluminium 13	28 **Si** silicon 14	31 **P** phosphorus 15	32 **S** sulfur 16	35.5 **Cl** chlorine 17	40 **Ar** argon 18
39 **K** potassium 19	40 **Ca** calcium 20	45 **Sc** scandium 21	48 **Ti** titanium 22	51 **V** vanadium 23	52 **Cr** chromium 24	55 **Mn** manganese 25	56 **Fe** iron 26	59 **Co** cobalt 27	59 **Ni** nickel 28	63.5 **Cu** copper 29	65 **Zn** zinc 30		70 **Ga** gallium 31	73 **Ge** germanium 32	75 **As** arsenic 33	79 **Se** selenium 34	80 **Br** bromine 35	84 **Kr** krypton 36
85 **Rb** rubidium 37	88 **Sr** strontium 38	89 **Y** yttrium 39	91 **Zr** zirconium 40	93 **Nb** niobium 41	96 **Mo** molybdenum 42	[98] **Tc** technetium 43	101 **Ru** ruthenium 44	103 **Rh** rhodium 45	106 **Pd** palladium 46	108 **Ag** silver 47	112 **Cd** cadmium 48		115 **In** indium 49	119 **Sn** tin 50	122 **Sb** antimony 51	128 **Te** tellurium 52	127 **I** iodine 53	131 **Xe** xenon 54
133 **Cs** caesium 55	137 **Ba** barium 56	139 **La*** lanthanum 57	178 **Hf** hafnium 72	181 **Ta** tantalum 73	184 **W** tungsten 74	186 **Re** rhenium 75	190 **Os** osmium 76	192 **Ir** iridium 77	195 **Pt** platinum 78	197 **Au** gold 79	201 **Hg** mercury 80		204 **Tl** thallium 81	207 **Pb** lead 82	209 **Bi** bismuth 83	[209] **Po** polonium 84	[210] **At** astatine 85	[222] **Rn** radon 86
[223] **Fr** francium 87	[226] **Ra** radium 88	[227] **Ac*** actinium 89	[261] **Rf** rutherfordium 104	[262] **Db** dubnium 105	[266] **Sg** seaborgium 106	[264] **Bh** bohrium 107	[277] **Hs** hassium 108	[268] **Mt** meitnerium 109	[271] **Ds** darmstadtium 110	[272] **Rg** roentgenium 111								

Elements with atomic numbers 112–116 have been reported but not fully authenticated

*The Lanthanides (atomic numbers 58–71) and the Actinides (atomic numbers 90–103) have been omitted.

Cu and **Cl** have not been rounded to the nearest whole number.

Index